Applications of Biotechnology for Sustainable Development

Kunal Mukhopadhyay
Ashish Sachan · Manish Kumar
Editors

Applications of Biotechnology for Sustainable Development

Springer

Editors
Kunal Mukhopadhyay
Department of Bio-Engineering
Birla Institute of Technology, Mesra
Ranchi, Jharkhand
India

Manish Kumar
Department of Bio-Engineering
Birla Institute of Technology, Mesra
Ranchi, Jharkhand
India

Ashish Sachan
Department of Bio-Engineering
Birla Institute of Technology, Mesra
Ranchi, Jharkhand
India

ISBN 978-981-10-5537-9 ISBN 978-981-10-5538-6 (eBook)
DOI 10.1007/978-981-10-5538-6

Library of Congress Control Number: 2017946955

Printed on acid-free paper

This Springer imprint is published by Springer Nature
The registered company is Springer Nature Singapore Pte Ltd.
The registered company address is: 152 Beach Road, #21-01/04 Gateway East, Singapore 189721, Singapore

Editorial Board for BSD

Organizing Committee for BSD2015

Patron: Dr. Manoj Kumar Mishra, Hon'ble Vice-Chancellor, BIT, Mesra
Chairman: Dr. Purnendu Ghosh, Executive Director, BISR, Jaipur
Convener: Dr. Kunal Mukhopadhyay, Professor & Head, Department of Bio-Engineering, BIT Mesra
Organizing Secretary: Dr. Ashish Sachan, Department of Bio-Engineering, BIT Mesra
Joint Secretaries: Dr. Vinod Kumar Nigam and Dr. Raju Poddar
Coordinators: Dr. Shashwati Ghosh Sachan and Dr. Sheela Chandra
Treasurer: Dr. Manish Kumar

Contents

About the Editors

Dr. Kunal Mukhopadhyay is Professor at the Department of Bio-Engineering, Birla Institute of Technology in Jharkhand, India. He obtained his Bachelor's degree from the Presidency College, Kolkata and M.Sc. and Ph.D. degrees from the University of Calcutta, India. He was selected for the Rockefeller Foundation Postdoctoral Fellowship in Rice Biotechnology Program and pursued his research at the University of Georgia, USA. He has worked extensively on wheat crop improvement. Dr. Mukhopadhyay has also worked in other areas of plant biotechnology, particularly genomics-driven metabolomics of medicinal plants like Guggul and Tulsi for the identification of key metabolites and regulation of genes involved in Guggulsterone and phenylpropanoid biosynthesis. He has completed seven research projects, published more than 30 research articles and contributed to three book chapters.

Dr. Ashish Sachan is Assistant Professor at the Department of Bio-Engineering, Birla Institute of Technology, Jharkhand, India. He obtained his M.Tech. and Ph.D. degrees from the Indian Institute of Technology Kharagpur (IIT Kharagpur), India. His main area of research is exploring the microorganisms for degradation of pollutants and production of value-added products. He is actively involved in research and has published more than 20 papers in international and national journals such as FUEL, Renewable & Sustainable Energy Reviews, Applied Microbiology & Biotechnology, Journals of Industrial Microbiology and Biotechnology, Annals of Microbiology and Letters in Microbiology. He is a life member of the Association of Microbiologists of India and the Biotech Research Society of India (BRSI).

Dr. Manish Kumar is Associate Professor at the Department of Bio-Engineering, Birla Institute of Technology, Jharkhand, India. He obtained his M.Sc. and Ph.D. degrees from Ranchi University. His main area of research is exploring the environment for degradation of pollutants and instrumentation. He is actively involved in research and has published more than 25 papers in international and national journals such as Planta, Plant Cell Reports, PLoS ONE and Gene. He is a life member of the Association of Microbiologists of India and the Indian Society for Technical Education.

Antibacterial Activity of *Euphorbia hirta* L.

Indu Kumari and R.K. Pandey

Abstract

Euphorbia hirta has been widely used by Tribal as traditional medicine in a treatment against infectious pathogens. *E. hirta* Linn. is a perennial herb belonging to the family Euphorbiaceae. It is a potent medicinal plant and has established its sedative and anxiolytic activity, analgesic, antipyretic, anti-inflammatory, antidepressant for blood pressure, antihypertensive, and antioxidant. Effect of extracts of different parts (Leaf, Bud, and Stem) of *E. hirta* L. using different solvents were examined using agar disc diffusion method against *Staphylococcus aureus*. Maximum in vitro inhibition of tested bacteria *S. aureus* was scored in methanol extracts of leaf or bud of *E. hirta* which offered same inhibition zone of 20 mm and Zone of Inhibition Area of 471.00 mm^2. A significant inhibition of bacteria *S. aureus* was found in aqueous or ethanol solvent extracts of *E. hirta*. A significant inhibition of 15 mm was obtained in ethanol extracts and 13 mm from aqueous extract of *E. hirta* against *S. aureus*.

Keywords

Antibacterial activity · Solvents · Extracts · *Staphylococcus aureus*

Introduction

Despite the existence of potent antibacterial agents, resistant or multi-resistant strains are continuously emerging, imposing the need for a continuous search and development of new drugs (Barbour et al. 2004; Machado et al. 2003; Rojas et al. 2003).

I. Kumari (✉)
Department of Botany, Nirmala College, Doranda, Ranchi, Jharkhand, India
e-mail: induraj0016@gmail.com

R.K. Pandey
University Department of Botany, Ranchi University, Ranchi, Jharkhand, India

K. Mukhopadhyay et al. (eds.), *Applications of Biotechnology for Sustainable Development*, DOI 10.1007/978-981-10-5538-6_1

At present, there is an urgent and continuous need of exploration and development of cheaper and effective new plant based drugs with better bioactive potential and least side effects. Consequently, this has led to the search for more effective antimicrobial agents among materials of plant origin, with the aim of discovering potentially useful active ingredients that can serve as source and template for the synthesis of new antimicrobial drugs (Pretorius et al. 2003; Moreillion et al. 2005).

Euphorbia hirta has been widely used by Tribal as traditional medicine in a treatment against infectious pathogens. *E. hirta* Linn. is a perennial herb belonging to the family Euphorbiaceae. It is a common weed referred to as 'garden spurge'. It is a slender-stemmed, annual hairy plant with many branches. The stem is hairy and the leaves are oblong, elliptical, acute, or subacute. Flowers are small, crowded and numerous seen together in thick cymes about 1 cm in diameter. The fruits are yellow in color, three-celled, hairy and have keeled capsules which are around 1–2 mm in diameter. They contain three four sided, brown, wrinkled, and angular seeds.

It is a potent medicinal plant and has established its sedative and anxiolytic activity, analgesic, antipyretic, anti-inflammatory, antidepressant for blood pressure, antihypertensive, and antioxidant. It is important in treating respiratory ailments, especially cough, coryza, bronchitis, and asthma.

Gram positive bacteria—*Staphylococcus aureus* can cause a range of illnesses, from minor skin infections, such as pimples, impetigo, boils (furuncles), cellulitis folliculitis, carbuncles, scalded skin syndrome, and abscesses, to life-threatening diseases such as pneumonia, meningitis, osteomyelitis, endocarditis, toxic shock syndrome (TSS), bacteremia, and sepsis. Its incidence ranges from skin, soft tissue, respiratory, bone, joint, endovascular to wound infections and has been selected for the present study.

Antibacterial activity of crude extracts of *E. hirta* against some few bacteria associated with enteric infections was studied by some scientist (El-Mahmood et al. 2009; Perumal et al. 2012; Ibrahim et al. 2012).

The purpose of the present study was to investigate the antibacterial activity of different parts of *E. hirta* using different solvents against disease causing bacteria—*S. aureus*.

Materials and Methods

Plant Material

Euphorbia hirta was collected as whole plant from different locations of Ranchi district of Jharkhand, India.

Preparation of the Extract

Collected Fresh leaves, buds, and stems were cleaned with water and dried in shade and pulverized into fine powdered substances by a grinding machine. Each 15 g of powder was transferred into conical flask. Then 150 mL of methanol was added in the flasks, closed by foil paper and placed on a shaker at 37 °C temperature for 72 h. The crude extract was then filtered by passing the extract through Whatman No. 1 filter paper and then concentrated. After complete solvent evaporation, extracts were weighed and stored in a refrigerator at 4 °C for further use. 500 mg of solvent residue was dissolved in 10 mL of solvents were used as the test extracts for antibacterial activity assay.

Test Bacteria

Pathogenic bacteria such as *S. aureus* was obtained from Birsa Agriculture University, Kanke, Ranchi, Jharkhand, India. The test bacterial species was maintained on nutrient agar media.

Antibacterial Activity

Antibacterial activity of leaf, bud or stem extract using different solvents: methanol, ethanol, dimethyl sulfoxide or aqueous were determined by disc diffusion method on nutrient agar medium. The filter paper discs of 5 mm diameter were prepared using Whatman No. 1 filter paper, soaked in extract. The discs dipped in respective solvent were used as negative controls. The petri-dishes were sterilized in hot air oven and nutrient agar medium was sterilized by autoclaving. This media was poured in the sterile petri-dishes. Test bacteria was spread on the solid plates with a sterile swab moistened with the bacterial suspension. Plates were incubated at 37 ° C for 24 h.

Antibacterial activities were evaluated by measuring zone of inhibition by using HiMedia zone scale in mm unit.

Results and Discussion

Effect of different solvent extract of different parts of *E. hirta* L. against *S. aureus* has been presented in Table 1. Different types of solvents were used such as methanol, ethanol, dimethyl sulfoxide or aqueous for studies of antibacterial activity of *E. hirta* against human pathogenic gram positive bacteria *S. aureus*. Among treatments, maximum in vitro inhibition of tested bacteria *S. aureus* was scored in methanol extract of leaf or bud of *E. hirta* which offered same Zone of inhibition of 20 mm and Zone of Inhibition Area of 471.00 mm^2.

Further, stem extract of *E. hirta* using methanol, leaf or bud extract of *E. hirta* using ethanol, bud, or stem extract of *E. hirta* using aqueous solvent and also leaf or bud extract of same plant in dimethyl sulfoxide were effective against *S. aureus*

Table 1 Antibacterial activity of *Euphorbia hirta* against *Staphylococcus aureus*

Solvent	Leaf		Bud		Stem	
	DIZ (mm)	ZIA (mm^2)	DIZ (mm)	ZIA (mm^2)	DIZ (mm)	ZIA (mm^2)
Methanol	20	471.00	20	471.00	15	294.38
Ethanol	15	294.38	15	294.38	11	181.34
Dimethyl sulfoxide	15	294.38	15	294.38	10	157.00
Aqueous	13	234.72	15	294.38	15	294.38

DIZ Diameter of zone of inhibition in millimeter scale
ZIA Zone of Inhibition Area in millimeter square

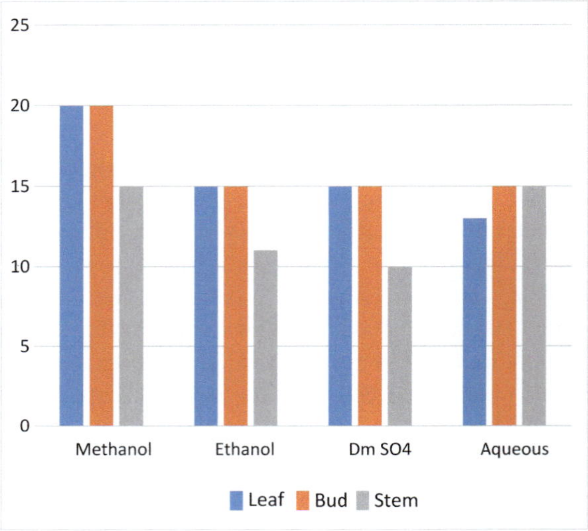

Fig. 1 Antibacterial activity of *Euphorbia hirta* against *Staphylococcus aureus*

which recorded same significant Zone of inhibition of 15 mm and Zone of Inhibition Area of 294.38 mm^2 (Fig. 1).

A significant inhibition zone of bacteria *S. aureus* was found in aqueous extracts of leaf of *E. hirta* showing 13 mm zone of inhibition. From Table 1, it is seen that the extract from different parts in different solvents such as methanol, ethanol, dimethyl sulfoxide or aqueous of *E. hirta* showed antibacterial activity against *S. aureus*. Methanol extract of *E. hirta* leaf showed the maximum degree of antibacterial activity properties. Crude ethanol extract of *E. hirta* stem powder or and Dimethyl sulfoxide extract of *E. hirta* stem produced the minimum 11 or 10 mm zone of inhibition respectively and Zone of Inhibition Area of 181.34 or 157.00 mm^2 respectively against *S. aureus*.

Conclusions

The result of this study showed that extracts of different parts of *E. hirta* have varied antibacterial activities against the tested organism. This suggests that the extracts of these plants are broad spectrum in their activities. The result of this study, confirms its use traditionally in treating antibacterial infections like dysentery, wound infection. These primary extracts open the possibility of finding new clinically effective antibacterial compounds.

Out of all the extracts from *E. hirta*, the methanol extract was the most active. It showed marked antibacterial activity against *S. aureus*. This may be due to the presence of alkaloids, tannins, saponins, and flavonoids, which are secondary metabolites of plants. These secondary metabolites are actually the defensive mechanisms of the plants against pathogens. However the present study of in vitro antibacterial evaluation of *E. hirta* forms a primary platform for further phytochemical and pharmacological studies to discover new antibiotic drugs.

References

Abubakar EMM (2009) Antibacterial activity of crude extracts of *Euphorbia hirta* against some bacteria associated with enteric infections. J Med Plants Res 3(7):498–505. ISSN 1996-0875

Barbour EK, Al Sharif M, Sagherian VK, Habre AN, Talhouk RS, Talhouk SN (2004) Screening of selected indigenous plants of Lebanon for antimicrobial acitivity. J Ethnopharmacol 93(1):1–7

Ibrahim TA, Adetuyi FO, Lola A (2012) Phytochemical screening and antibacterial activity of *Sida acuta* and *Euphorbia hirta*. J Appl Phytotechnol Environ Sanitation 1(3):113–119. ISSN 2088–6586

Machado TB, Pinto AV, Pinto MCFR, Leal ICR, Silva MG, Amaral ACF, Kuster RM, Netto-dosSantos KR (2003) In-vitro activity of Brazilian medicinal plants, naturally occurring naphthoqinones and their analogues, against methicillin–resistant *Staphylococcus aureus*. Int J Antimicrob Agents 21:279–284

Moreillion P, Que YA, Glauser MP (2005) *Staphylococcus aureus* (including staphyloccal toxic shock). In: Mandell GL, Bennett JE, Dolin R (eds) Principles and practice of infectious diseases, 6th edn, vol 2. Churchill Livingstone, Pennsylvania, pp 2333–2339

Perumal S, Pillai S, Cai LW, Mahmud R, Ramanathan S (2012) Determination of minimum inhibitory concentration of *Euphorbia hirta* (L.) extracts by tetrazolium microplate assay. J Nat Prod 5:68–76. ISSN 0974-5211

Pretorius JC, Magama S, Zietsman PC (2003) Growth inhibition of plant pathogenic bacteria and fungi by extracts from selected South African plant species. S Afr J Bot 20:188–192

Rojas R, Bustamante B, Bauer J (2003) Antimicrobial activity of selected Peruvian medicinal plants. J Ethanopharm 88:199–204

Molecular Characterization of *Anogeissus acuminata* Genotypes Employing RAPD Markers

Sanjay Singh, Kanchan Kumari, Shweta Chaturvedi, Nutan Pandey and Ashley Varghese

Abstract

Anogeissus acuminata (Roxb. ex DC.) Wall. ex Guill. & Perr.** is commonly known as button tree and locally Phasi/Pasi. It is the specific kind of tree species with spiritual and cognitive environs of Odisha as its timber is useful in famous Ratha Yatra to make chariot wheels of Lord Jagannatha. Due to its potency and efficient qualities; *Anogeissus* timber is of implicit value, including its medicinal and economical values. The present work was aimed at studying the genetic diversity and relationship existing among the genotypes and discrimination at the inter-specific and intra-specific level, employing RAPD markers. Using 22 primers, 166 distinct bands were generated out of which 106 bands (66.63%) were found to be polymorphic. The amplicons obtained after PCR were run on 1.5% gel and was visualized under gel documentation system. The timber availability has become a challenge during the festival every year. So, it is necessary to conserve this species for future generations. Thus, genetic diversity and molecular characterization assessment through DNA-based molecular markers may be reliable and authentic for the species. In plant breeding programmes the most diverse genotypes shall be used in the hybridization programme.

Keywords

A. acuminata · Genetic diversity · Species conservation · RAPD markers

---

S. Singh (✉) · K. Kumari · S. Chaturvedi · N. Pandey · A. Varghese
Institute of Forest Productivity, Ranchi Gumla,
National Highway - 23, Lalgutwa, Ranchi 835303, Jharkhand, India
e-mail: sanjaysingh@icfre.org

K. Mukhopadhyay et al. (eds.), *Applications of Biotechnology for Sustainable Development*, DOI 10.1007/978-981-10-5538-6_2

Introduction

Biotechnology is the manipulation of living organisms or their components to produce useful and/or usually commercial products, viz. pest resistant crops, new bacterial strains or novel pharmaceuticals. Highly instructive DNA markers have been developed which helped immensely in the identification of genetic polymorphism. In molecular biology techniques, generally polymerase chain reaction (PCR)-based random amplified polymorphic DNA (RAPD) markers are used.

A study conducted by IFP Ranchi has found the no. of Phasi trees is decreasing to a significant extent due to selective logging of promising trees for Ratha Yatra purpose. The timber availability has become a challenge during the festival every year. So it is necessary to conserve this species for future generations. It is also necessary to check whether sufficient genetic diversity is present in natural populations or not. Genetic diversity and molecular characterization assessment through DNA-based molecular markers may be reliable and authentic for the species. To study the genetic diversity and relationship existing among the various genotypes of *Anogeissus acuminata* through RAPD markers is the objective of our present study.

Materials and Methods

Sample Collection

Fresh young leaves of 8 different germplasms of *A. acuminata* were collected and assembled from botanical garden of IFP.

DNA Extraction, Quantification and Gel Electrophoresis

Young leaf tissues that are visually clean and unaffected by pathogens of each genotypes were collected for genomic DNA isolation and analysis. DNA extractions following CTAB method with slight modification was done to remove the interfering polyphenolic compounds and other polysaccharides. Quantification of DNA was accomplished by UV spectrophotometer (Biophotometer plus, Genetics) and by analyzing the DNA on 0.8% agarose gel. DNA was diluted in TE buffer to a concentration of 50 ng/μl for use in PCR amplification.

Twenty-two desultory decamer oligonucleotide primers (RPI series, Bangalore GeNei) were used for RAPD-based PCR amplification of genomic DNA. Table 2 contains information regarding primer sequence. The PCR reaction mix included 50 ng/μl template DNA, 1X Taq Buffer A (Tris with 15 mM $MgCl_2$), deoxynucleotide triphosphates (dNTPs) each of 0.2 mM (Banglore GeNei). In a Thermal cycler namely GeneAmp* PCR 9700; all PCR amplification reactions were performed. The PCR conditions included: the first step as denaturation at 94 °C for 4 min, followed by 40 cycles of 1 min at 94 °C, annealing temperature of 1 min at

30–50 °C (annealing temperature was chosen based on gradient) and extension at 72 °C for 2 min. A final extension of 7 min at 72 °C was used for primer extension. The amplicons obtained were separated on 1.5% Agarose gel in 1X TAE buffer containing 5 μg/ml ethidium bromide. The size of fragments amplified was assessed by comparing to the ΦX174/Hae III digest DNA Ladder (Banglore GeNei) and gel was photographed on Gel Documentation System (Syngene).

Amplicons Assay

Amplified product for RAPD analysis were scored on the basis of presence (taken as 1) or absence (taken as 0) of band for each primer. Obvious and clear bands/fragments were considered as potential RAPD markers. The percentage of polymorphism and monomorphism for each primer of eight genotypes of *A. acuminata* was determined. Without considering the band intensity, the observed bands were taken significantly thus avoiding taxonomic resemblance. The basic measures for population genetic analysis were calculated by subjecting the data to POPGENE Software.

Results and Discussion

Evaluation using genetic diversity and variation is prerequisite for genetic improvement of important plant species to enhance biomass/production, increase oil yield, develop stress tolerance and use of elite accessions/genotypes. The present disquisition was conducted to ascertain the superior genotypes of *Anogeissus acuminata*, a tree species with multiple utilities, through molecular characterization employing the available RAPD markers.

Gel Scoring and Data Analysis

To determine the purity of a DNA solution a comparison of optical density value at various wavelengths was employed. In case of pure DNA, the absorbance ratio

Table 1 Data obtained for quantification of DNA by biophotometer

S. No.	Population	Quantity (ng/μl)	A_{230}	A_{260}	A_{280}	A_{340}	$A_{260/280}$	$A_{260/230}$
1	Pop1	2.9	0.056	0.059	0.047	0.017	1.25	1.05
2	Pop2	2.4	0.050	0.047	0.042	0.016	1.13	0.96
3	Pop3	1.4	0.026	0.028	0.029	0.012	0.97	0.08
4	Pop4	13.4	0.259	0.267	0.269	0.249	0.99	1.03
5	Pop5	1.0	0.020	0.020	0.018	0.006	1.11	0.99
6	Pop6	0.5	0.009	0.009	0.012	0.002	0.73	1.00
7	Pop7	0.4	0.008	0.008	0.012	0.005	0.69	1.00
8	Pop8	63.3	1.361	1.267	1.244	1.173	1.02	0.93

Table 2 Amplified DNA bands and polymorphism generated in *A. acuminata* genotype using 22 RAPD markers

S. No	Primer	Sequence of the primer	Total no. of amplified products	No. of polymorphic products	No. of monomorphic products	Polymorphisms (%)	Tm (°C)	Annealing temperatures (°C)
1	RPI01	AAAGCTGCGG	7	3	4	42.85	32	35
2	RPI02	AACGCGTCGG	5	4	1	80	34	35
3	RPI03	AAGCGACCTG	2	2	0	100	32	40
4	RPI04	AATCGCGCTG	5	2	3	40	32	45
5	RPI05	AATCGGGCTG	4	3	1	75	32	40
6	RPI06	ACACACGCTG	8	6	2	75	32	35
7	RPI07	ACATCGCCCA	8	5	3	62.8	32	45
8	RPI08	ACCACCCACC	10	7	3	70	34	50
9	RPI09	ACCGCCTATG	4	2	2	50	32	35
10	RPI10	AGCATGAGCG	7	4	3	57.14	32	40
11	RPI11	ACGGAAGTGG	6	5	1	83.33	32	30
12	RPI12	ACGGCAACCT	8	3	5	37.5	32	40
13	RPI13	ACGGCAAGGA	15	9	6	60	32	40
14	RPI14	ACTTCGCCAC	9	7	2	77.77	32	40
15	RPI15	ACCTGAAGCC	6	5	1	83.33	32	40
16	RPI16	AGGCGGCAAG	8	8	0	100	34	40
17	RPI17	AGGCGGGAAC	11	8	3	72.72	34	40
18	RPI18	AGGCTGTGTC	11	6	5	54.54	32	40
19	RPI20	AGTCCGCCTC	9	3	6	33.33	34	40
20	RPI21	CACGAACCTC	7	4	3	57.14	32	45
21	RPI24	CCAGCCGAAC	4	3	1	75	34	45
22	RPI25	GAGCGCCTTC	12	7	5	58.33	34	40
Total			166	106	60			
Average			7.54	4.8	2.72	63.66		

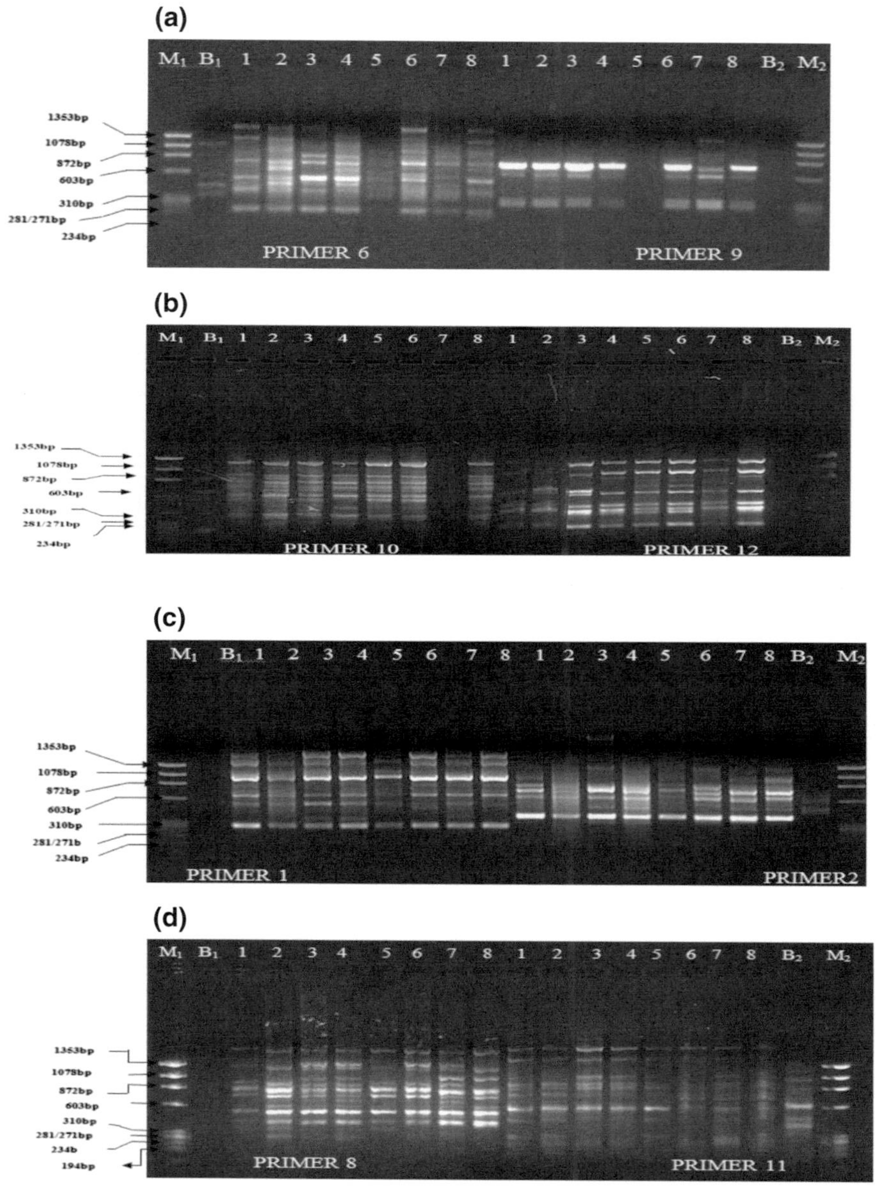

Fig. 1 **a–f** RAPD profile of eight genotypes of *Anogeissus acuminata* using the primer RPI#01, RPI#02, RPI#06, RPI#09, RPI#10, RPI#12, RPI#08, RPI#11, RPI#21, RPI# 24, RPI# 25, RPI#15 and RPI#16. *M* represent molecular weight and size of marker (ΦX174/Hae III digest DNA Ladder). The *numbers* represent different genotypes. M_1 and M_2—100 base pair ladder (B_1 and B_2—blank)

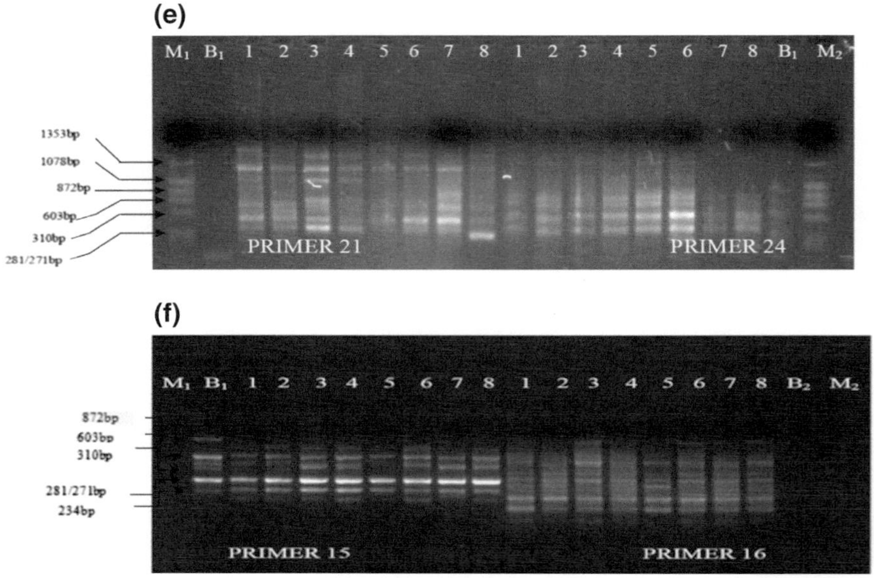

Fig. 1 (continued)

$A_{260/280}$ should be near to 1.8. Generally elevated ratio indicates the presence of RNA. $A_{260/280}$ ratio less than 1.8 often signals the phenol or protein contamination. Alternately, $A_{260/230}$ ratio greater than 0.5 indicates protein and phenol contamination. Phenol contamination is detected with a peak at 270 nm and an $A_{260/280}$ ratio 1.2. DNA having phenol uncontamination will have $A_{260/280}$ of around 2. Absorption at 230 nm can be caused by contamination by phenolate ion, thiocyanate and other organic compounds. Absorption at 330 nm and higher indicates particulate contamination causing light scattering. Thus we can say that in above table population 1 has phenol contamination (Table 1).

Genetic relationships among the eight genotypes have been carried out using RAPD markers. The particulars of the primers yielding amplified DNA bands, their nucleotide sequence, amplicons number, monomorphic/polymorphic bands count, unique bands count and amplified bands range are mounted in Table 2.

Band patterns produced by some of the RAPD primers have been shown in Table 2. The use of RAPD primer generated a total of 166 distinct band/fragments out of which 106 (63.85%) bands were found to be polymorphic and 60 bands were monomorphic. Primers were ranged from 2 to 15 wherein each primer produced an average of 8 bands. The primer RPI16 generated maximum number (8) of polymorphic fragments while RPI04 produced minimum number of polymorphic bands that is 2. The percentage of polymorphism ranged from 33.33 to 100%. Primer RPI03 and RPI16 showed 100% polymorphism where as RPI20 showed 33.33% polymorphism. In their ability to discover polymorphism, various primers showed

Table 3 Nei's genetic identity (above diagonal) and genetic distance (below diagonal)

Pop ID	1	2	3	4	5	6	7	8
1	****	0.7725	0.6826	0.5988	0.5928	0.6886	0.6527	0.6826
2	0.2582	****	0.7305	0.6467	0.6287	0.7365	0.7126	0.7066
3	0.3818	0.3140	****	0.7964	0.6707	0.7904	0.7784	0.7246
4	0.5128	0.4359	0.2276	****	0.6707	0.7425	0.6946	0.7246
5	0.5229	0.4640	0.3995	0.3995	****	0.6407	0.6766	0.6946
6	0.3731	0.3058	0.2352	0.2977	0.4452	****	0.7365	0.8263
7	0.4266	0.3389	0.2505	0.3644	0.3906	0.3058	****	0.7425
8	0.3818	0.3473	0.3222	0.3222	0.3644	0.1907	0.2977	****

variation but all the 22 primers were effective in determining variation among the *acuminata* genotypes (Fig. 1).

The amplified products were analyzed using the software POPGENE. Genetic Distance Nei's (1973) analysis of genetic distance and identity was used to study the genetic distance between the eight populations. Table 3 shows that the highest genetic identity exists between the pop 7 and pop 8, i.e 0.8263 and maximum genetic distance exists between pop 1 and pop 5, i.e. 0.5928. In plant breeding programmes, the most diverse genotypes shall be used in the hybridization programme.

According to the dendrogram (Fig. 2) based on Nei's analysis (1973), the whole genotype was divided into two clusters, cluster I containing pop 1, pop 2, pop 3, pop 4, pop 6, pop 7, pop 8 and the cluster II contain pop 5. Pop 5 stood separate while rests of the population were grouped together. Cluster I was sub-divided 2 groups pop 1 and pop 2 grouped together. Separation from the rest and pop 3, pop 4, pop 6, pop 7 and pop 8 were in another group. This group was further divided into groups where pop 7 remain separately and the rest pop 3, pop 4, pop 6 and pop 8 formed another group. Now the group containing pop 3, pop 4, pop 6 and pop 8 were further divided into two groups where pop 3 and pop 4 remain together and pop 6 and pop 8 remain together.

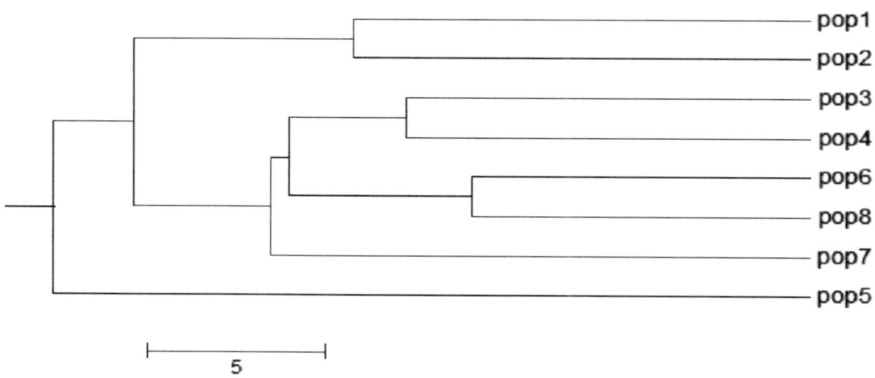

Fig. 2 Dendrogram obtained from RAPD analysis based on Nei's (1973) genetic distance

According to the dendrogram, pop 1 and pop 2 were highly similar to each other with respect to pop 3, pop 4, pop 6, pop 8, pop 7 and pop 5 which seem to be recently separated. Similarly pop 3 and pop 4 are more similar to each other than the pop 6, pop 8, pop 7 as they are separated in the earlier stage. Pop 6 and pop 8 are also highly similar with respect to pop 3 and pop 4 after undergoing further division. The genotype of plant population shows the interaction of a lot of various processes such as in the origin and evolution of the species (e.g. distribution curve and/or species isolation), mutation types, genetic drift in a population, mating system of species, gene migration, inheritance and natural selection. The above factors may be responsible for complex genetic structure within a population. The sustainable development of any plant population relates to its genetic diversity. In RAPD assay polymorphism is seen due to various types of mutations viz. insertion, deletion or substitution of one or more nitrogen bases inside the priming site sequence. However, it is apparent that RAPD markers can be used in a reproducible manner for detection of inter-specific and intra-specific differences.

Reference

Nei M (1973) Analysis of gene diversity in sub divided populations. Proc Natl Acad Sci U S A 70:3321–3323

An Efficient Protocol for Plant Regeneration of *Phlogacanthus thyrsiflorus* Nees: An Important Medicinal Shrub

Shweta Singh, Madhuparna Banerjee and Manish Kumar

Abstract

Phlogacanthus thyrsiflorus Nees (Vernacular name: Rambasak) is a gregarious shrub belonging to family Acanthaceae, widely used in the northeastern states of India for its medicinal value. The objective of the present work was to develop a rapid and efficient protocol for in vitro propagation of *P. thyrsiflorus* Nees. The efficacy of shoot multiplication was tested by inoculating nodal explants in Murashige and Skoog's (MS) medium alone and supplemented with different concentrations of BA, Kn, or combination of BA with NAA and BA with adenine sulphate along with 1 mg l^{-1} citric acid to minimise the phenolic exudation. Highest rate of multiplication with an average numbers of 13 shoots per explant was observed in MS medium supplemented with 2 mg l^{-1} BA, 50 mg l^{-1} adenine sulphate and 1 mg l^{-1} citric acid after six weeks of culture. Rooting of excised shootlets was found sufficient in half strength MS medium supplemented with 0.5 mg IBA. Within 10 days of inoculation, an average of 8 roots per explant was observed. The plants were hardened in poly tunnel for eight days followed by secondary hardening in green house for twenty days with 95% plant survival rate.

Keywords

Phlogacanthus thyrsiflorus Nees · Plant regeneration · MS medium

---

S. Singh · M. Kumar (✉)
Department of Bio-Engineering, Birla Institute of Technology, Mesra, Ranchi 835215, Jharkhand, India
e-mail: manish@bitmesra.ac.in

M. Banerjee
College of Biotechnology, Birsa Agricultural University, Kanke, Ranchi 834006, Jharkhand, India

K. Mukhopadhyay et al. (eds.), *Applications of Biotechnology for Sustainable Development*, DOI 10.1007/978-981-10-5538-6_3

Introduction

Phlogacanthus thyrsiflorus Nees commonly known as Ram Vasak or Titaphul is an evergreen shrub of medicinal importance belonging to family Acanthaceae. Acanthaceae is a large sub-tropical family of about 250 genera and 2500 species (Panigrahi and Patel 2014). *P. thyrsiflorus* Nees is found in subtropical Himalayas, Upper Gangetic plain, Bihar, Bengal and Assam (Khare 2007). In the northeastern states, *P. thyrsiflorus* is considered therapeutically superior over its allied species, *Adhatoda vasica* prescribed as an expectorant, (quinazoline alkaloids being active principle in *A. vasica* leaf). The whole plant is of high medicinal value, used by the tribes of northeast India since long for various ailments. Its different aerial parts are used for the remedy of cough, asthma, allergy, fever, jaundice, dysentery, tuberculosis, malarial fever and rheumatism (Kalita and Bora 2008; Jaiswal 2010). In vitro analgesic, hypolipidaemic and hypoglycemic potentials have also been reported (Mukherjee et al. 2009; Ilham et al. 2012; Chakravarty and Chandra 2012).

In vitro multiplied medicinal plants acts as a starting material for the biotransformation experimentation aimed at the enrichment of biologically active compounds. The objective of the present study was to establish a reproducible protocol for mass multiplication and preservation of this medicinally important plant resource. Seeds of *P. thyrsiflorus* are nonviable and are propagated by vegetative means. This technique provides a continuous supply of plant materials, which can help exploit the therapeutic properties of this plant species and eliminate the need for harvesting specimens from the wild. Also, this protocol could be applied to genetic manipulation through transformation techniques.

Materials and Methods

Plant Material Collection and Shoot Proliferation

P. thyrsiflorus plant was obtained from the nursery of College of Biotechnology, Birsa Agricultural University, Ranchi. The material was identified by Botanical Survey of India (No. CNH/Tech. II/2015/4/243) and the voucher specimen was deposited at the Department of Pharmaceutical Sciences and Technology, Birla Institute of Technology, Mesra, Ranchi (voucher specimen No. PhSc-0463/13-14). The young, actively growing nodal segments (2–3 cm) from lateral branches were collected and washed thoroughly under running tap water and later surface sterilised in 0.1% bavistin (systemic fungicide) solution (w/v) for 15 min followed by rinsing 3–4 times with distilled water. Finally, under laminar air flow cabinet the nodal explants were surface sterilised using 0.1% $HgCl_2$ (w/v) for 5 min followed by thorough rinsing with autoclaved distilled water. These sterile nodal explants were inoculated aseptically in MS medium (Murashige and Skoog 1962) solidified with agar (0.75%) and supplemented with different phytohormones along with

1 mg l^{-1} citric acid to minimise the phenolic exudation. Phytohormones selected for shoot multiplication were BA alone and supplemented with different concentrations of BA, Kn or combination of BA with NAA and BA with adenine sulphate along with 1 mg l^{-1} citric acid. pH of the media was adjusted between 5.6 and 5.8 and autoclaved at 121.6 °C and 15 PSI pressure for 16 min. Cultures were maintained in the growth chamber with 16 h/8 h (light/dark) photoperiod at 25 ± 2 °C with light intensity of 25 µmol m^{-2} s^{-1} by cool-white fluorescent lamps.

Root Induction and Acclimatisation

Rooting of in vitro multiplied shootlets was induced in half strength MS medium supplemented with IBA in different concentrations. The plantlets thus generated were hardened primarily in portray containing vermicompost: coco peat (1:1) inside a poly tunnel for eight days followed by secondary hardening for twenty days in poly bags containing FYM: garden soil: sand (1:1:1) inside green house.

Statistical Analysis

The experiments were conducted with a minimum of four replicates and were repeated three times. Results obtained were expressed as mean ± standard deviation.

Results and Discussion

In the present study, plants of *P. thyrsiflorus* were successfully regenerated from its nodal segments (Fig. 1). Surface sterilised nodal explants when inoculated in MS medium supplemented with different concentrations and combinations of different cytokinins alone (BA: 0.5–3.0 mg l^{-1}, Kn: 1.0–3.0 mg l^{-1}) or in combinations of BA and NAA (1.0 + 0.5, 2.0 + 0.5 and 3.0 + 0.5 mg l^{-1}) and in combinations of BA and adenine sulphate (1.0 + 25, 2.0 + 25, 3.0 + 25, 1.0 + 50, 2.0 + 50, 3.0 + 50 mg l^{-1}) along with citric acid (1 mg l^{-1}) resulted in varied response in shoot proliferation, number of shoots and shoot lengths (Table 1). Shoot multiplication of nodal explants in MS medium supplemented with BA or kinetin alone or BA and NAA yielded poor results. Addition of 0.5 mg l^{-1} NAA to BA (in different concentrations) resulted in an increase in shoot bud breaking response and shoot length but number of shoots was not much different when compared with the inoculation in MS medium with BA(in different concentrations) alone. Inoculation in media containing Kinetin failed to give response better than in BA alone. BA in combination with adenine sulphate resulted increase in both numbers and length of shootlets. Best response of 94.44 ± 0.10% was obtained from explants inoculated in MS medium supplemented with 2 mg l^{-1} BA + 50 mg l^{-1} adenine sulphate and 1 mg l^{-1} citric acid after six weeks, with mean numbers of shoots per explant being

Fig. 1 Micropropagation of *P. thyrsiflorus*: **a** explant after 2 weeks of inoculation in MS medium + 2 mg l^{-1} BA + 50 mg l^{-1} adenine sulphate + 1 mg l^{-1} citric acid, **b** 4 weeks old, **c** 6 weeks old, **d** rooting in multiplied shoots in half MS medium + 0.5 mg l^{-1} IBA, **e** primary hardening, **f** secondary hardening

Table 1 Effect of different concentrations and combinations of phytohormones on shoot multiplications of *P. thyrsiflorus* Nees nodal explants after six weeks of inoculation

Hormone concentration (mg l^{-1})	Response (%)	Mean no. of shoots	Average shoot length (mm)
BA			
0.5	15.00 ± 0.13	1.20 ± 0.76	11.50 ± 0.65
1.0	32.78 ± 0.08	2.68 ± 0.16	18.76 ± 0.86
2.0	46.66 ± 0.06	3.49 ± 0.22	20.72 ± 0.48
3.0	61.67 ± 0.13	4.56 ± 0.21	31.62 ± 0.76
Kn			
1.0	20.56 ± 0.04	1.40 ± 0.53	11.87 ± 0.58
2.0	27.22 ± 0.11	2.04 ± 0.39	15.70 ± 0.48
3.0	32.78 ± 0.08	2.56 ± 0.21	20.89 ± 0.39
BA + NAA			
1.0 + 0.5	55.00 ± 0.18	3.05 ± 0.42	25.51 ± 0.43
2.0 + 0.5	60.55 ± 0.19	3.81 ± 0.20	32.13 ± 0.29
3.0 + 0.5	68.33 ± 0.16	4.82 ± 0.16	35.85 ± 0.87
BA + adenine sulphate			
1.0 + 25	61.67 ± 0.13	4.65 ± 0.96	31.04 ± 0.80
2.0 + 25	68.33 ± 0.16	7.78 ± 0.41	41.63 ± 0.26
3.0 + 25	79.45 ± 0.04	9.13 ± 0.55	45.68 ± 0.61
1.0 + 50	82.20 ± 0.17	11.34 ± 0.64	43.70 ± 0.26
2.0 + 50	**94.44 ± 0.10**	**13.43 ± 0.60**	**53.65 ± 0.49**
3.0 + 50	87.78 ± 0.11	9.90 ± 0.23	49.07 ± 0.90

Data are represented as mean ± standard deviation

13.43 ± 0.60 and average length of shoots 5.37 ± 0.49 cm (Table 1). This combination of PGRs suited best for obtaining large numbers of shoots with a good plant height. Use of adenine sulphate with BA resulting in higher rate of multiplication has been documented earlier in many other plants (Muruganantham et al. 2002; Nandgopal and Kumari 2006; Naaz et al. 2014). Rooting was obtained from shootlets inoculated in MS medium supplemented with different concentrations of IBA. Within 10 days of inoculation, 0.5 mg l^{-1} IBA in half strength MS medium resulted in an average of 8.0 roots per shootlets. This concentration was found sufficient in regeneration of *P. thyrsiflorus* plants. Higher concentrations of IBA (1–10 mg l^{-1}) resulted in over rooting. MS medium alone was taken as control, both for shoot multiplication and root induction. Single protocol for plant regeneration has been published till date (Hassan et al. 2011) in which a maximum of 12.4 ± 0.66 shoots were obtained in MS medium supplemented with 1.0 mg l^{-1} BA + 0.5 mg l^{-1} NAA after repeated subcultures. Rooting was best observed in half strength MS medium supplemented with 0.5 mg l^{-1} IBA + 0.5 mg l^{-1} NAA. The present work is an easily reproducible protocol for in vitro multiplication of *P. thyrsiflorus* with a high number of plant regeneration rate.

Conclusion

Micropropagation of medicinal plant can act as starting material for many of the metabolomics work through molecular biology techniques. The above is an easily reproducible protocol for mass multiplication of *P. thyrsiflorus*. Also, aseptically multiplied shootlets can be used for biotransformation studies.

Acknowledgements The authors are grateful to the Centre of Excellence, Department of Bio-Engineering under Technical Education Quality Improvement Program-II (Grant No. NPIU/TEQIP II/FIN/31/158) for financial support and Department of Agriculture, Government of Jharkhand (Grant No. 5/B.K.V/Misc/12/2001) for providing infrastructure development fund.

References

Chakravarty S, Chandra KJ (2012) Antihyperglycaemic effect of flower of *Phlogacanthus thyrsiflorus* Nees on streptozotocin induced diabetic mice. Asian Pac J Trop Biomed S1357–S1361

Hassan AKMS, Begum N, Sultana R, Khatun R (2011) In vitro shoot proliferation and plant regeneration of *Phlogacanthus thyrsiflorus* nees. A rare medicinal shrub of Bangladesh. Plant Tissue Cult Biotech 21(2):135–141

Ilham S, Ali MS, Hasan CM, Kaisar MA, Bachar SC (2012) Antinociceptive and hypoglycemic activity of methanolic extract of *Phlogacanthus thyrsiflorus* Nees. Asian J Pharma Clin Res 5(2):15–18

Jaiswal V (2010) Culture and ethnobotany of Jaintia tribal community of Meghalaya, northeast India—a mini review. Indian J Tradit Knowl 9(1):38–44

Kalita D, Bora RL (2008) Some folk medicines from Lakhimpur district, Assam. Indian J Tradit Knowl 7(3):414–416

Khare CP (2007) Indian medicinal plant, an illustrated dictionary, vol 1. Springer, Berlin, p 478

Mukherjee A, Chaliha M, Das S (2009) Study of analgesic activity of ethanol extract of *Phlogacanthus thyrsiflorus* on experimental animal models. Bangladesh J Pharmacol 4:147–149

Murashige T, Skoog F (1962) A revised medium for rapid growth and bioassays with tissue cultures. Physiol Plant 15:473–497

Muruganantham M, Ganapathy A, Selvaraj N, Anand RP, Vasudevan A, Vengadesan G (2002) Adenine sulphate and L-glutamine enhance multiple shoot induction from cotyledon explants of Melon (*Cucumis melo* L. cv. Swarna). Cucurbit Genet Coop Rep 25:22–24

Naaz A, Shahzad A, Anis M (2014) Effect of adenine sulphate interaction on growth and development of shoot regeneration and inhibition of shoot tip necrosis under in vitro condition in adult *Syzygium cumini* L.—a multipurpose tree. Appl Biochem Biotechnol 173:90–102

Nandgopal S, Kumari BDR (2006) Adenine sulphate induced high frequency shoot organogenesis in callus and in vitro flowering of *Cichorium intybus* L. cv. Focus—a potent medicinal plant. Acta Agric Slov 87(2):415–425

Panigrahi J, Patel AN (2014) Micro propagation of *Adhatoda beddomei* using nodal explants. Eur Acad Res 2(9):12194–12204

Cloning, Evolutionary Relationship and Microarray-Based Expression Analysis of WRKY Transcription Factors in Wheat (*Triticum aestivum* L.)

Lopamudra Satapathy and Kunal Mukhopadhyay

Abstract

WRKY, a plant-specific transcription factor family, have important roles in pathogen defense, abiotic stress and phytohormone signalling. Members of this family contain approximately 60 amino acids containing at least one WRKY domain such as highly conserved WRKYGQK peptide and a zinc finger motif. In this study, gDNA of three *TaWRKY* genes were cloned and found to encode complete protein sequences. Phylogenetic tree revealed that these genes belonged to class II and III WRKY based on the number of WRKY domains and the pattern of zinc finger structures. In order to investigate the putative role of cis-acting elements which are responsible for gene expression under developmental and abiotic stress, in silico promoter analysis was performed for the cloned genes using PLACE database. Microarray-based developmental expression analysis was also performed for the identified *TaWRKY* genes in order to determine their levels of expression across different stages of development from seedling to anthesis through heat map analysis. Expression analysis of *WRKY* genes suggests their role during normal plant development in a tissue-specific manner. The results presented here will help in the design of experiments for functional validation of the precise role of selected WRKYs in wheat development and abiotic stress responses.

Keywords

Wheat · Transcription factor · WRKY · Abiotic stress

L. Satapathy · K. Mukhopadhyay (✉)
Department of Bio-Engineering, Birla Institute of Technology,
Mesra, Ranchi 835215, Jharkhand, India
e-mail: kmukhopadhyay@bitmesra.ac.in

K. Mukhopadhyay et al. (eds.), *Applications of Biotechnology for Sustainable Development*, DOI 10.1007/978-981-10-5538-6_4

Introduction

Wheat (*Triticum aestivum* L.) provides 40–60% of calories in human diet. The demand of wheat is expected to increase by 60% in the year 2050, whereas the production is estimated to decrease 29% due to environmental stress factors (Manickavelu et al. 2012). Hence, comprehension of molecular mechanisms such as transcription factors (TFs) underlying the stress tolerance in wheat has a great importance for improvement of wheat cultivars (Okay et al. 2014). Members of WRKY contain approximately 60 amino acids containing at least one WRKY domain such as highly conserved WRKYGQK peptide and a zinc finger motif. Based on the number of WRKY domains and the type of zinc finger like motif, WRKY TFs have been classified into three classes (Rushton et al. 2010). WRKY proteins with two WRKY domains were classified as class I. Class II WRKY is marked by presence of one WRKY domain with C_2H_2 zinc finger structure. The class III of WRKY proteins carries a single domain and differs from class I and II in its altered C_2HC zinc finger motif ($C–X_7–C–X_{23}–H–X–C$) (Ulker and Somssich 2004; Wei et al. 2012). In this study, gDNA cloning and promoter analysis was performed for WRKY sequences identified by Satapathy et al. (2014). Also, microarray-based expression analysis was performed for the identified wheat WRKY TFs.

Materials and Methods

DNA Isolation, Primer Designing, Molecular Cloning and Sequencing

Wheat cultivar HD2329 ± *Lr*28 gene was used for DNA extraction from the whole plant 12 days after germination using Qiagen DNeasy Plant Mini Kit (Qiagen). Primers were designed for (TaWRKY56, TaWRKY60 and TaWRKY62) using Primer Express 2.0. DNA amplifications were performed with annealing temperature for TaWRKY56, TaWRKY60 and TaWRKY62 as 62, 60 and 50 °C respectively. The amplified product was gel purified using Thermos Scientific silica bead DNA gel extraction kit, cloned in PTZ57R/T (Thermo Scientific) via T/A cloning. Plasmids from three independent clones were sequenced from both ends using M13 universal primer. Neighbour joining phylogenetic tree was constructed using protein sequences of the TaWRKY sequences with 1000 bootstrap replicates.

Promoter Analysis

For promoter analysis, WRKY sequences 1000 bp upstream of the initiation codon were retrieved and searched using PLACE database to identify cis-regulatory elements.

Microarray-Based Expression Analysis of *TaWRKY* Genes Using GEO Database

Six rust-related microarray-based experimental data were downloaded from NCBI-GEO (National Centre for Biotechnology Information—Gene Expression Omnibus) and uploaded onto the sample set. The experiment IDs included GSE6227, GSE9915, GSE31753, GSE31756, GSE31761 and GSE32151. The expression patterns of the wheat WRKY genes were determined using Genevestigator software (Hruz et al. 2008). Expression data were hierarchically clustered based on Euclidean distance in sample with log2 transformed values.

Results and Discussion

Isolation of gDNA, Amplification and Cloning

Genomic DNA isolated from the leaves of HD2329 + *Lr28* (Fig. 1) were of high molecular weight (~30 Kb). 200 ng of gDNA was used as template for PCR amplification with designed primers for *WRKY* genes (Table 1). Amplicon length of

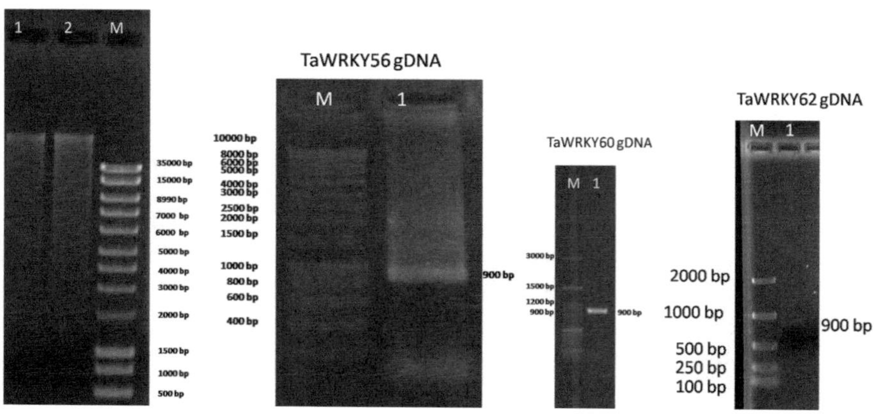

Fig. 1 Agarose gel profile of genomic DNA isolated from HD2329 + *Lr28*, Lane 1, 2 and Lane M: molecular weight marker (DNA Supermix ladder, Merck). Amplification of genomic DNA clones of *TaWRKY56*, *60* and *62*

Table 1 Primer sequences used for amplification of TaWRKY56, 60 and 62 gDNAs

S. No.	*TaWRKY* gene name with accession ID	Primer sequence 5′-3′ (forward-top; reverse bottom)	Amplicon length (bp)
1.	*TaWRKY56* (HX157194)	ATGTCCTCCTACTCGTCCCTCC	655
		CATGTCAATTGTCAGATATCACGCT	
2.	*TaWRKY60* (HX157222)	CATGTCAATTGTCAGATATCACGC	599
		GAGCAGATCGGTGGGTATGC	
3.	*TaWRKY62* (BE404156)	AGCGCATGAGGAAGGTGGAT	600
		GCATGCGTGTAGATGGGAAAA	

Fig. 2 Neighbour-joining phylogenetic tree of TaWRKY proteins with 1000 bootstrap replications

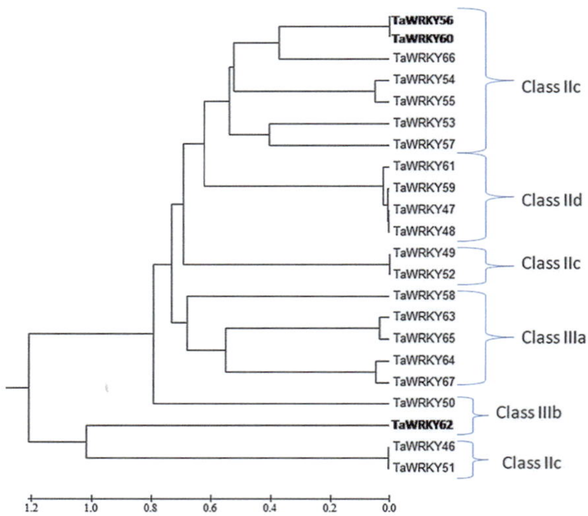

900 bp was generated for gDNA clones of TaWRKY56, 60 and 62 (Fig. 1). Plasmids from three independent clones were sequenced using Sanger dideoxy sequencing.

Analysis of Cloned Sequences

VecScreen was used to remove the vector portion in the forward and reverse sequence. The forward and reverse-complement orientations were compared and the best possible contigs were assembled using Sequencher 5.1. The sequence obtained was compared with those available at Gene Bank database using BLASTN at NCBI. *TaWRKY60* and *62* were found to be intronless, whereas *TaWRKY56* showed introns at position (283-60 and 544-407). Phylogenetic tree of the cloned sequences (Fig. 2) along with the newly identified 19 TaWRKY proteins depict that TaWRKY56 and TaWRKY60 belong to Class IIc (WRKYGKK domain) and TaWRKY62 (WRKYGEK domain) belong to class IIIb. Although the phylogenetic tree showed some divergences, as members of class IIc were found in separate clades. This might be due to gene duplication or recombination resulting in domain rearrangements events (Wei et al. 2012).

Cis-Regulatory Element Analysis

TATA box (core promoter element) and light-responsive elements such as TCT-motif, TCCC motif, Sp1 motif, G-box, CATT-box, Box-4 were found in the promoter regions of the cloned genes suggesting their involvement in photoperiodic control regulation during flowering. Hormone specific expression such as CCGTCC-box, TCA element, TGACG motif, CCGTCC box, CGTCA-motif, ABRE has been observed in all these genes. Circadian control motif was present in *TaWRKY62* only. Meristem specific regulation such as CAT box and fungal elicitor responsive elements (Box W1) were present in *TaWRKY56* and *60*. ARE element which is responsible for anaerobic induction was present in *TaWRKY56* only and GC motif responsible for anoxic specific inducibility was present in all three genes. Presence of O_2 motif and LTR motif confers stress response and zein metabolism respectively in wheat.

Microarray-Based Expression Analysis of *TaWRKY* Genes Using GEO Database

Developmental expression analysis was performed for 14 *TaWRKY* genes to determine their levels of expression in rust condition across different stages of

Fig. 3 Development patterns expression in *TaWRKY* genes

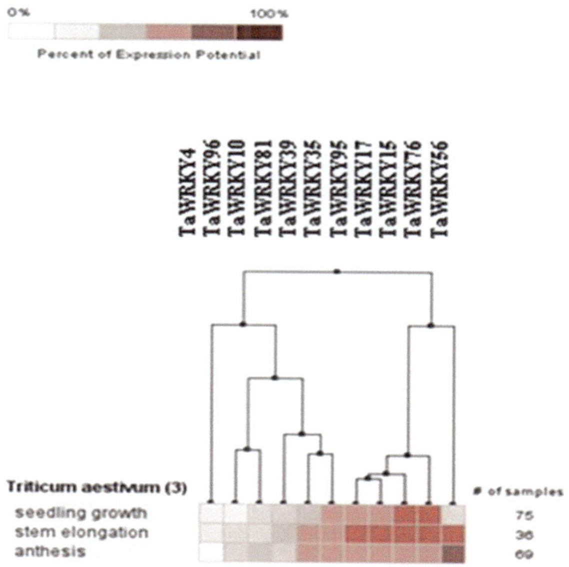

development i.e. from seedling to anthesis through heat map analysis (Fig. 3). *TaWRKY15*, *17*, *76* and *95* showed highest expression in the stem elongation stage whereas, TaWRKY56 showed highest expression in anthesis stage. TaWRKY4, 10, 81 and 96 showed downregulation in the developmental patterns.

Conclusion

In this study 3 *TaWRKY* genes, i.e. *TaWRKY56*, *60* and *62* were cloned. Their phylogenetic analysis revealed that TaWRKY56 and TaWRKY60 belonged to Class IIc and TaWRKY62 belonged to class IIIb WRKY. *Cis*-regulatory elements distribution among *TaWRKY56*, *60* and *62* revealed that they are associated with vital physiological processes including response to light, hormone response, meristem specific expression, stress response and endosperm expression. Developmental-based expression analysis performed for the *TaWRKY* genes through heat map analysis revealed that *TaWRKY56* showed highest expression during anthesis stage.

Acknowledgements This work was supported by Department of Biotechnology, Government of India (Grant No. BT/PR6037/AGR/02/308/05), BTISNet SubDIC (BT/BI/04/065/04), Department of Agriculture, Government of Jharkhand (5/B.K.V/Misc/12/2001) and CoE-TEQIP-II (Grant No. NPIU/TEQIP II/FIN/31/158). L.S. is grateful to Department of Science and Technology—INSPIRE (Fellowship/2011/318) for Ph.D. fellowship.

References

Hruz T, Laule O, Szabo G, Wessendorp F, Bleuler S, Oertle L, Widmayer P, Gruissem W, Zimmermann P (2008) Genevestigator V3: a reference expression database for the meta-analysis of transcriptomes. Adv Bioinf 420747. doi:10.1155/2008/420747

Manickavelu A, Kawaura K, Oishi K, Shin IT, Kohara Y, Yahiaoui N, Keller B, Abe R, Suzuki A, Nagayama T, Yano K, Ogihara Y (2012) Comprehensive functional analyses of expressed sequence tags in common wheat (*Triticum aestivum*). DNA Res 19:165–177

Okay S, Derelli E, Unver T (2014) Transcriptome-wide identification of bread wheat WRKY transcription factors in response to drought stress. Mol Genet Genomics 289:765–781

Rushton PJ, Somssich IE, Ringler P, Shen QJ (2010) WRKY transcription factors. Trends Plant Sci 15:247–258

Satapathy L, Singh D, Ranjan P, Kumar D, Kumar M, Prabhu KV, Mukhopadhyay K (2014) Transcriptome-wide analysis of WRKY transcription factors in wheat and their leaf rust responsive expression profiling. Mol Genet Genomics 289:1289–1306

Ulker B, Somssich IE (2004) WRKY transcription factors: from DNA binding towards biological function. Curr Opin Plant Biol 7:491–498

Wei KF, Chen J, Chen YF, Wu LJ, Xie DX (2012) Molecular phylogenetic and expression analysis of the complete WRKY transcription factor family in maize. DNA Res 19:153–164

Leaf Rust Responsive Expression Analysis of TIFY Transcription Factor Family in Wheat (*Triticum aestivum* L.)

Poonam Singh and Kunal Mukhopadhyay

Abstract

Bread wheat (*Triticum aestivum* L.) is one of the major staple food crops grown globally and provides food calories and proteins (http:/www.faostat.fao.org). Rusts are important fungal diseases identified in wheat worldwide. Leaf rust can cause decrease in yield ranging from 20 to 95%. TIFY is a plant-specific transcription factor family which shows differential expression during leaf rust disease. It is defined by the presence of the 36 amino acid TIFY domain (previously ZIM) and is classified into four different subfamilies TIFY, PPD, JAZ and ZML, based on the variations in their domain architectures. Some TIFY genes have been highly upregulated during leaf rust infection. Pearson coefficient showed difference in expression patterns of wheat TIFY genes in four different SAGE (serial analysis gene expression) libraries prepared from pathogen and mock inoculated wheat near-isogenic lines. Coexpression studies have also been carried out to identify which of these genes are expressed together during leaf rust pathogenesis. TaTIFY genes which are upregulated during leaf rust might be plays important roles to provide resistance to the plants. The difference in Pearson correlation coefficient indicates that susceptible and resistant-NILs utilize a different set of TaTIFY genes to counter leaf rust pathogenesis. The genes which are clustered together in coexpression network might be expressed together during leaf rust pathogenesis to provide resistance to the plant.

Keywords

Leaf rust · Wheat · TIFY · Transcription factors · SAGE · WGCNA

P. Singh · K. Mukhopadhyay (✉)
Department of Bio-Engineering, Birla Institute of Technology, Mesra, Ranchi 835215, Jharkhand, India
e-mail: kmukhopadhyay@bitmesra.ac.in

K. Mukhopadhyay et al. (eds.), *Applications of Biotechnology for Sustainable Development*, DOI 10.1007/978-981-10-5538-6_5

Introduction

Bread wheat (*Triticum aestivum* L.) is one of the most popular food crops grown all over the world. It provides major part of food calories and proteins (http:/www.faostat.fao.org). Rust disease popularly known as 'polio of agriculture' is the major limitation to its worldwide cultivation. Three important rust diseases in wheat comprise leaf, stem, and stripe rust. Leaf rust, caused by *Puccinia triticina* Erikss., comprises major rust disease in wheat. Leaf rust can cause losses in yield ranging from 20 to 95% (Kolmer et al. 2009). The extensive distribution of these fungi, its capacity to form new races which can attack resistant cultivars, capability of its spores to move distant places, and potential to germinate swiftly under wide environmental conditions make the situation more pathetic for plant breeders to develop new variety. These fungi cause diseases on a number of angiosperm and gymnosperm plants (Kolmer et al. 2009). Expression of genes largely depends on the transcription factors and so to understand the mechanisms of gene expression under leaf rust, the first step would be the analysis of transcription factors.

Rusts are one of the most important fungal diseases of wheat all over the world. The reason behind this is their extensive distribution, the ability to form new races which can attack resistant cultivars, the ability of spores to move distant places, and potential to germinate swiftly under wide environmental conditions. The rust fungi are widespread plant pathogens that cause diseases on many angiosperm and gymnosperm plants. Stripe, leaf, and stem rust are three important rust diseases affecting wheat plant. Leaf rust and stripe (yellow) rust, caused by *P. triticina* Erikss. and *Puccinia striiformis* Westend. f. sp. tritici Erikss., respectively, are major rust diseases in wheat. Leaf rust can cause losses in yield ranging between 25 and 90% and stripe rust is responsible for 100% yield loss, but often ranges between 10 and 70%. The wheat rust fungi are obligate parasites that require living host to complete their life cycle and are biotrophs due to the method of extracting nutrients from living host cells. The entire disease cycle is complex, involving two plant hosts (wheat and an alternate host) and several different spore types to complete the life cycle. Most rust fungi are highly specialized pathogens specific to certain host species, due to the characteristics of parasitism involved. Rust fungi can be widely distributed over wide geographic areas by wind-borne basidiospores, aeciospores, and urediniospores and are often genetically diverse pathotypes differentiated by virulence/avirulence to differential host genotypes. Although fungicides can control rust diseases, costs and potential for negative environmental impacts are most important drawbacks to this strategy. The use of resistant varieties is the most proficient and cost-effective way to combat these diseases.

Stress gene induction occurs mainly at transcriptional level, and regulating their spatio-temporal expression is an important part of the plant stress response. Transcription factors (TFs) are proteins that act together with regulators of transcription factors, to employ or block RNA polymerases to bind to the template DNA. Plant genomes allocate approximately 7% of their coding sequence to TFs, which is responsible for the complexity of transcriptional regulation. TFs are basically proteins that bind to specific DNA sequences, generally to a motif in the target gene promoter, and control the transcription of the target gene. Plants being sessile organisms have to face a wide range of environmental stresses. Signaling cascades governing developmental and stress-related switches unite at the gene expression level. Researchers have suggested that transcriptional regulation might play important roles in plants as compared with animals, given a large number of TF-coding genes in plant genomes, ranging from 6 to 10%. Regulation of gene expression is crucial for various fundamental processes in plants, such as growth, development, differentiation, metabolic regulation, and adaptation to different environmental conditions. Transcription is the first step in the expression of any gene and plays central role in the regulation of the gene expression. Transcription appears to be controlled by various transcription factors that mediate the effects of intracellular and extracellular signals. Therefore, their analysis is essential for an understanding the underlaying mechanisms of gene expression.

The TIFY family was earlier known as ZIM (Zinc finger protein expressed in Inflorescence Meristem). This gene family is divided into four groups based on phylogenetic and structural analyses: TIFY (Threonine, Isoleucine, Phenylalanine, Tyrosine), JAZ (Jasmonate TIFY domain), PPD (PEAPOD) and ZML (zinc finger protein expressed in inflorescence meristem, TIFY-like) proteins (Bai et al. 2011; Vanholme et al. 2007).

Materials and Methods

Plant and Pathogen Material, Seedling Inoculation

A pair of near-isogenic lines (NILs) was used in this study involving HD2329 and HD2329 + *Lr28* genes (conferring rust seedling resistance). The wheat NILs HD2329 ± *Lr28* was used to isolate RNA for TIFY gene characterization studies. *Lr 28* effective against all pathotypes of *P. triticina* in India (Bipinraj et al., 2011).

The leaf rust pathogen race *P. triticina* selected as the experimental pathogen, it is reported to be the major leaf rust pathogen in all parts of Indian subcontinent. The race is avirulent against seedling leaf rust resistance gene *Lr28*. This pathogen race was used for inoculating the seedlings of the above wheat near-isogenic lines. The seedling avirulence/virulence formula for this race is P *Lr9*, *Lr18*, *Lr19*, *Lr24*, *Lr25*, *Lr 28*, *Lr 29*, *Lr 32*, *Lr 41*, *Lr 45/p*, *Lr 1*, *Lr 2*, *Lr3*, *Lr10*, *Lr11*, *Lr12*, *Lr13*, *Lr14*, *Lr15*, *Lr16*, *Lr17*, *Lr18*, *Lr20*, *Lr22*, *Lr23*, *Lr26*, *Lr27+Lr31*, *Lr33*, *Lr34*, *Lr36*, *Lr37*, *Lr42*, *Lr43*, *Lr44*, *Lr46*, *Lr48*, *Lr49*.

The pathogen inoculums were prepared by mixing the urediniospores of *P. triticina* pathotype 77-5 and talcum powder at a ratio of 1:1. The inoculum was then applied gently on leaves of the HD2329 ± *Lr28*. Mock inoculation of the plant was done with talcum powder. The plants were then placed high humidity condition of >90% for 24 hpi (hours post inoculation) in the dark to aid the infection. After that pots were placed in the normal growth chamber [22 °C, day-time; 14 °C, night time, relative humidity (80%)] at the National Phytotron Facility, IARI, New Delhi.

RNA Isolation, SAGE Library Preparation and Next Generation Sequencing

Leaf tissue from 20 seedlings, of mock and pathogen inoculated HD2329 ± *Lr28* were procured at 24 hpi and stored in liquid nitrogen. Total RNA was isolated using TRIREAGENT (Molecular Research Centre, Inc, USA) as per manufacturer's instructions. The integrity of total RNA was confirmed using Agilent Bioanalyser 2100. Four serial analysis of gene expression (SAGE) libraries were prepared from the obtained RNA samples [coded as: (i) HD2329 mock inoculated: S-M, (ii) HD2329 pathogen inoculated: S-PI, (iii) HD2329 + Lr28 mock inoculated: R-M and (iv) HD2329 + Lr28 pathogen inoculated: R-PI] using SOLiD–SAGE kit (Applied Biosystems, CA, USA) as per manufacturer's instruction and sequencing was done using oligonucleotide ligation and detection (SOLiD) technique at Bay Zoltán Foundation of Applied Research, Institute of Plant Genomics, Human Biotechnology and Bioenergy, Zagreb, Hungary.

Expression Study of TIFY TFs in Wheat Under Leaf Rust Infection

Multiple database searches were performed to identify TIFY genes in wheat. The TIFY TF datasets of *Zea mays*, *Sorghum bicolor*, *Brachypodium distachyon* and *Oryza sativa* subspecies (both *indica* and *japonica*) were downloaded from Plant Transcription Factor Database (http://plntfdb.bio.uni-postsdam) and Gramineae Transcription Factor DataBase (http://gramineaetfdb.psc.riken.jp/). The retrieved protein sequences were used as queries for similarity search with wheat expressed sequence tags (ESTs) using TBLASTN at NCBI, with an e-value cutoff of 10 to predict GRAS TF genes in wheat.

Expression values for GRAS TFs in wheat upon leaf rust infection were extracted from the four SOLiD–SAGE library using CLC Genomics Workbench 6.5. The differential expression of the GRAS transcription factors among the wheat NILs in response to leaf rust pathogenesis was determined using log2 transformed values and represented using cluster analysis, scatter plot, and heat map. Expression data were hierarchically clustered on the basis of Euclidean distance in the sample with log2 transformed values.

Results and Discussion

Total RNA Isolation and cDNA Conversion

Total RNA extracted from all the collected samples were quantified and resolved on formaldehyde agarose gels to check the integrity of isolated RNAs (Fig. 1).

Expression Analysis of Identified TaTIFY TFs in Wheat Under Leaf Rust Stress

In order to identify the differentially expressed TaTIFY genes in response to leaf rust pathogenesis, the TIFY genes were mapped to the reads of four SOLiD–SAGE libraries using CLC Genomics Workbench 6.5. The number of times each TIFY gene mapped on the four SOLiD SAGE libraries represented an index for

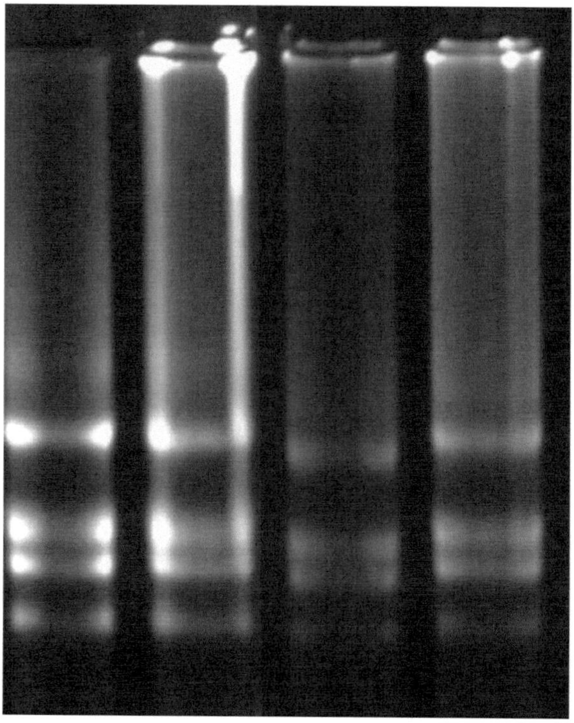

Fig. 1 RNA isolated from mock and inoculated samples of wheat plant

Table 1 Read counts obtained after mapping 23 *TIFY* genes to the four SOLiD SAGE libraries using CLC Genomics Workbench 6.5 and their corresponding fold change in different libraries

Transcription factor	S-M	S-PI	R-M	R-PI	S-M versus S-PI fold change	R-PI versus S-PI fold change	R-PI versus R-M fold change
TaTIFY1	6	11	54	265	−1.11	−48.41	−3.65
TaTIFY2	11	5	6	24	3.61	−9.61	−2.96
TaTIFY3	2	21	2	14	−6.39	−1.23	−4.82
TaTIFY4	18	10	4	12	2.95	−2.4	−2.22
TaTIFY5	25	4	51	7	1.72	−4	8.58
TaTIFY6	104	5	29	63	34.13	−24.42	−1.56
TaTIFY7	21	8	18	13	4.3	−3.42	1.86
TaTIFY8	14	34	11	9	−1.47	1.88	1.64
TaTIFY9	5	25	3	4	−3.04	3.12	1.01
TaTIFY10	3	12	9	2	−2.43	1.99	4.04
TaTIFY11	4	3	7	14	2.18	−9.34	−1.84
TaTIFY12	12	4	10	2	−2.43	1	6.73
TaTIFY13	10	3	7	2	5.47	1.33	4.71
TaTIFY14	14	24	7	3	−1.04	−3.99	3.14
TaTIFY15	11	3	3	2	6.01	−2	1.34
TaTIFY16	4	12	4	3	−1.82	1.99	1.79
TaTIFY17	33	18	24	47	3	−5.45	−1.51
TaTIFY18	9	3	3	3	4.92	−2	1.34
TaTIFY19	14	2	20	4	11.48	−4	1.79
TaTIFY20	6	5	2	6	1.96	−2.4	−1.51
TaTIFY21	7	4	11	2	2.87	−1.5	1.34
TaTIFY22	9	10	7	11	1.47	−2.2	6.73
TaTIFY23	2	64	2	9	−19.49	3.55	−2.22

estimation of their relative abundance (Table 1). The pair wise experiments were conducted between the four different libraries using differentially expressed reads with log2 transformed values to obtain hierarchical clustering and were displayed as heat map (Fig. 2).

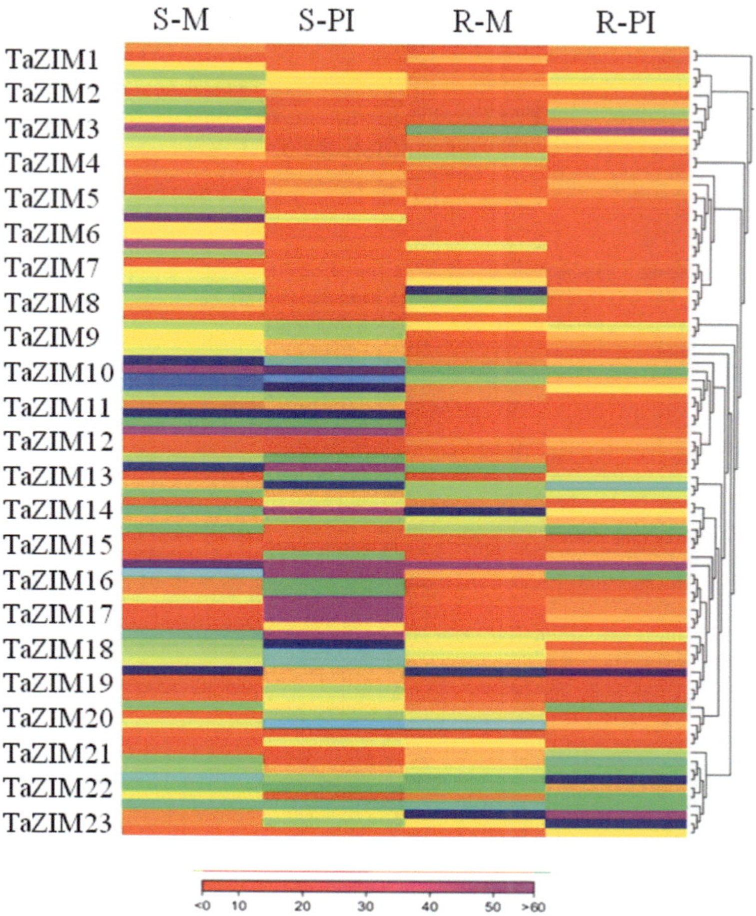

Fig. 2 Heat map illustration of differentially expressed TaGRAS genes from wheat near-isogenic lines showing major changes in gene expression in SAGE libraries corresponding to S-M (susceptible mock), S-PI (susceptible pathogen inoculated), R-M (resistant mock) and R-PI (resistant pathogen inoculated)

Conclusion

To the best of our knowledge this is the first report of TIFY TFs in response to leaf rust pathogenesis in wheat. The up- and down-regulation of TIFY TFs have been studied during leaf rust pathogenesis. In the susceptible plants the maximum level of expression after pathogen infection was observed in five TaTIFY genes (TaTIFY6, 7, 13, 15, and 19) and minimum level of expression was observed in six TaTIFY genes (TaTIFY3, 9, 10, 12, 16, and 23) while in the case of resistant plants maximum level of expression was obtained in six TaTIFY genes (5, 12, 13, 14, and

22) and minimum level of expression was obtained in 5 TaTIFY genes (TaTIFY1, 2, 3, 4, and 23). In both susceptible- and resistant plants, the TaTIFY15 is observed as highly expressed gene, whereas TaTIFY3 and TaTIFY13 are found to have reduced expression. TaTIFY12 and 3 genes can be considered as genes having minimal role in infection. Therefore, rest of the TaTIFY genes might have some role during *Puccinia* infection which could be revealed by conducting further studies.

Acknowledgements The author's acknowledge CoE-TEQIP-II (Grant no. NPIU/TEQIP II/FIN/31/158) for financial support, BTISNet SubDIC (BT/BI/04/065/04) and Department of Agriculture, Government of Jharkhand (5/B.K.V/Misc/12/2001) for providing infrastructure development fund.

References

Bai Y, Meng Y, Huang D, Qi Y, Chen M (2011) Origin and evolutionary analysis of the plant-specific TIFY transcription factor family. Genomics 98:128–136

Bipinraj A, Honrao B, Prashar M, Bhardwaj S, Rao S et al (2011) Validation and identification of molecular markers linked to the leaf rust resistance gene Lr28 in wheat. J Appl Genetics 52:171–175

Kolmer JA, Ordonez ME, Groth JV (2009) The rust fungi. In: Encyclopedia of life sciences, Wiley Ltd., Chichester, UK, pp 1–8

Vanholme B, Grunewald W, Bateman A, Kohchi T, Gheysen G (2007) The TIFY family previously known as ZIM. Trends Plant Sci 12:239–244

A Correlation Study Between Drug Resistance and Plasmid Profiling

Monalisa Padhan and Smaranika Pattnaik

Abstract

This work was intended to correlate the drug resistance among the clinical as well as nosocomial bacterial strains with plasmid profile of respective strains. The antibiogram study of thirty-seven Gram-negative bacterial strains was made by using nine numbers of commonly prescribed antibiotics (Ampicillin, Penicillin, Vancomycin, Streptomycin, Tetracycline, Ciprofloxacin, Levofloxacin, Ofloxacin and Chloramphenicol) using the 'Disc diffusion' and 'Agar cup' method. The degree of resistance was measured qualitatively in terms of highly resistant (100–75% resistance), moderately resistant (75–50%) and resistant (50–25%) based upon a respective number of bacterial strain resistance towards the antibiotics under study. Further, these resistant strains were subjected to plasmid profile analysis by using Hi-purA TM plasmid DNA purification kit (Hi-Media, Mumbai). The gel electrophoresis was illustrating about the presence of plasmid bands in a range of 1–7 in number. The molecular weights of plasmids were in the range from 800 bp to ≤ 10 kbp. However, some of the resistant strains were devoid of plasmid bands. A correlation regression analysis was made between the number of antibiotics to which the bacteria were resistant (both clinical and nosocomial isolates) with the number of plasmid bands. Hence, this study is apprehensive about the correlation between acquisition of plasmid bands and degree of drug resistance.

M. Padhan · S. Pattnaik (✉)
Laboratory of Medical Microbiology, School of Life Sciences, Sambalpur University, Jyoti Vihar, Burla 768019, Odisha, India
e-mail: smaranika2010@gmail.com

M. Padhan
e-mail: monalisa.padhan@gmail.com

K. Mukhopadhyay et al. (eds.), *Applications of Biotechnology for Sustainable Development*, DOI 10.1007/978-981-10-5538-6_6

Keywords
Nosocomial · Clinical · Drug resistance · Plasmid profile · Correlation regression analysis

Introduction

Bacterial resistance to commonly used antibiotics is a threat to public health throughout the world. Multiple antibiotic resistances in bacteria are most commonly associated with the presence of plasmids which are the potential vehicles for drug-resistant genes. Transmission of resistance genes from normally more virulent pathogenic species to non-pathogenic organisms is very common with the animal and human intestinal tract microflora (Uma et al. 2009). Veer Surendra Sai Institute of Medical Sciences and Research (VIMSR) is a Premier health hub in Western Odisha. The people of this region are heavily dependent upon this healthcare institute. But the budding problem in drug therapy is due to the prevalence of drug-resistant strains belonging to MDR (multi-drug resistance) as well as XDR (extended drug resistance) categories. In this context, it is obligatory to screen for underlying mechanism behind the drug resistance phenomena conferred by the above-said organisms. Plasmids (the extra-chromosomal DNA) are the active carriers of drug-resistant genes among the bacterial strains because of their ability to shuttle between the strains. Therefore, this study aims at screening for drug resistance among the MDR/XDR bacterial strains isolated from clinical samples as well as the nosocomial environment and the respective plasmid profiling (Adnan et al. 2015; Odeyemi et al. 2015).

Materials and Methods

Isolation of Bacterial Strains

The bacterial strains used in this study were isolated from clinical samples as well as from hospital environment of VIMSR, Burla, Odisha using 'Swabbing method'. The swabs containing the bacterial inoculums were spread onto the selective media namely EMB agar (Himedia, Mumbai). The plates were incubated in bacteriological incubators for a period of 12–14 h and observations were made based upon the appearance of colonies with green metallic sheen that were identified as strains (MS+) of *E. coli*. However, the colonies growing on EMB agar plates without metallic sheen (GNR) were also included in this study.

Antibiotic Susceptibility Testing

The "Sensitivity test" (by following Disc diffusion and Agar cup) for antibiotics Ampicillin(10 μg/disc), Chloramphenicol (30 μg), Tetracycline (30 μg), Ciprofloxacin (5 μg), Levofloxacin (5 μg), Ofloxacin (5 μg), Penicillin-G (2 units/disc), Streptomycin (25 μg) and Vancomycin (10 μg) against respective bacterial strains were carried out by using standard protocol of Indian Pharmacopoeia and results were correlated with Kirby and Bauer's (Bauer et al. 1966) interpretative chart. Based upon "Antibiogram" the isolated bacterial strains were categorized as Resistant (R), Sensitive (S) or Intermediately sensitive (IS).

Extraction of Plasmid DNA

The plasmid DNA of all the 37 number of bacterial strains were extracted by using "HiPurA™ Plasmid DNA Miniprep Purification Kit". The plasmid profile was studied (Pattnaik et al. 2013) by Agarose gel electrophoresis and bands were visualized by UV transilluminator.

Correlation Study by Using Regression Analysis

A correlation study between the degree of drug resistance (the number of antibiotics to which the bacteria were resistant) and acquisition of number plasmid bands (plasmid profile) was carried out by applying correlation and regression analytical tools.

Results and Discussion

Isolation of Samples

A total number of 37 bacterial strains were isolated, out of which 13 were clinical and rest 24 were nosocomial site strains. In addition, 16 number of strains were showing green metallic sheen (MS) and the rest 21 number of bacterial strains did not show any metallic sheen (GNR).

Antibiotic Susceptibility Testing

The degree of resistance was measured qualitatively in terms of highly resistant (100–75% resistance), moderately resistant (75–50%) and resistant (50–25%) based upon number of bacterial strains showing resistance towards the antibiotics, included in this study. The results obtained from this 'Antibiotic susceptibility test' are depicted in Figs. 1, 2 and 3.

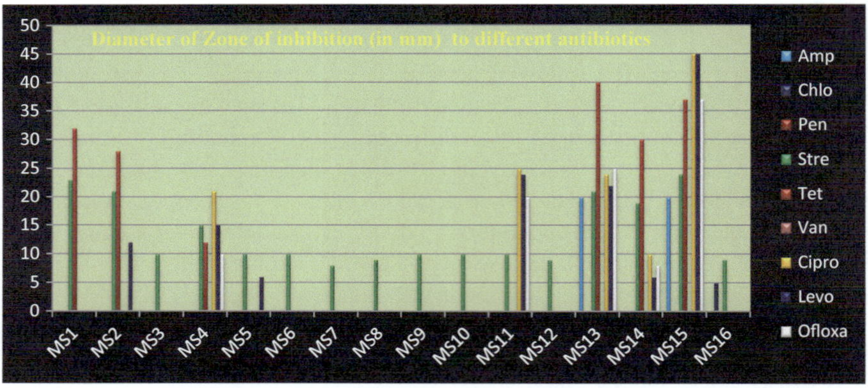

Fig. 1 Depicting the antibiogram of MS bacterial strains

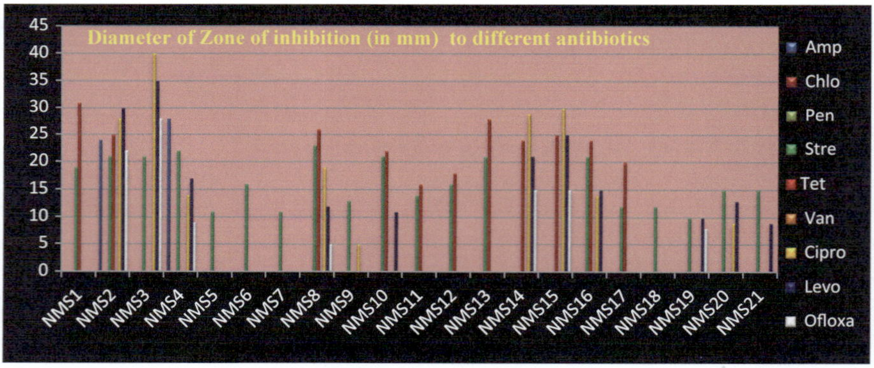

Fig. 2 Depicting the antibiogram of GNR bacterial strains

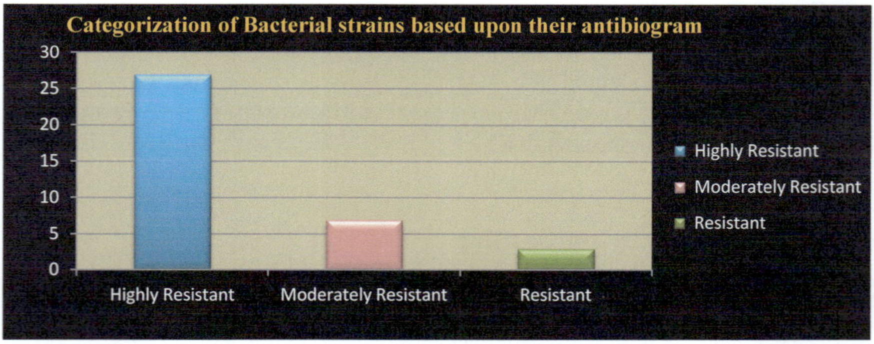

Fig. 3 Depicting the categorization of bacterial strains based upon their antibiogram

Extraction of Plasmid DNA

From the experiments on the screening for the presence of plasmid, bands in drug-resistant strains are depicted in Figs. 4 and 5. However, seven number of bacterial strains were not harboured with any plasmids. Hence, the plasmid profile of bacterial strains were observed and depicted in a graphical sketch in percentages (Figs. 6, 7 and 8).

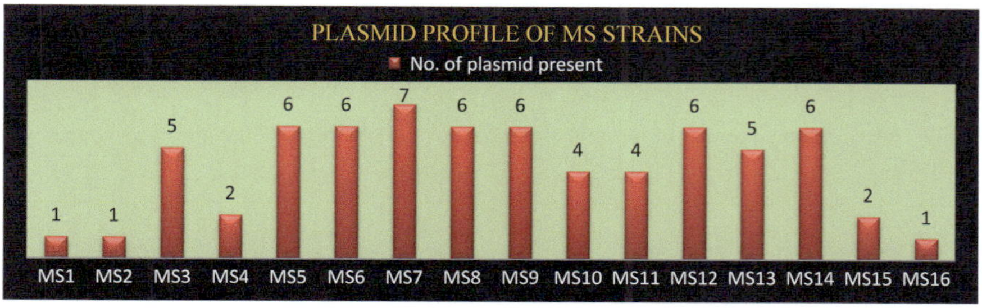

Fig. 4 Depicting the number of plasmids present in MS+ strains

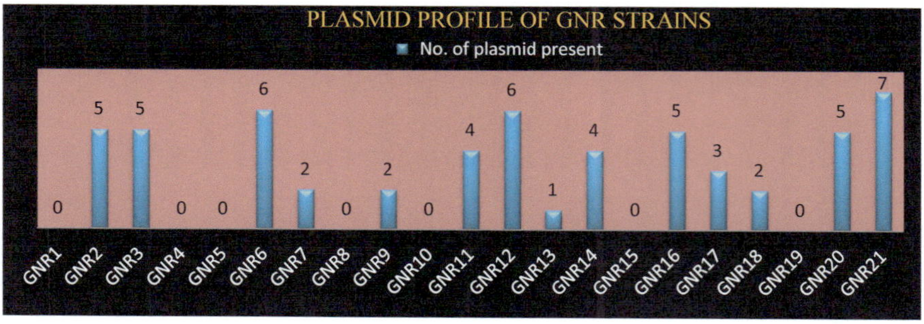

Fig. 5 Depicting the number of plasmids present in GNR strains

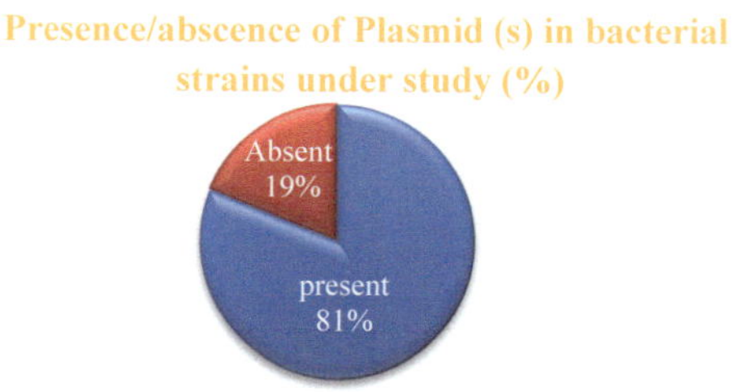

Fig. 6 Depicting the presence of plasmids in percentage

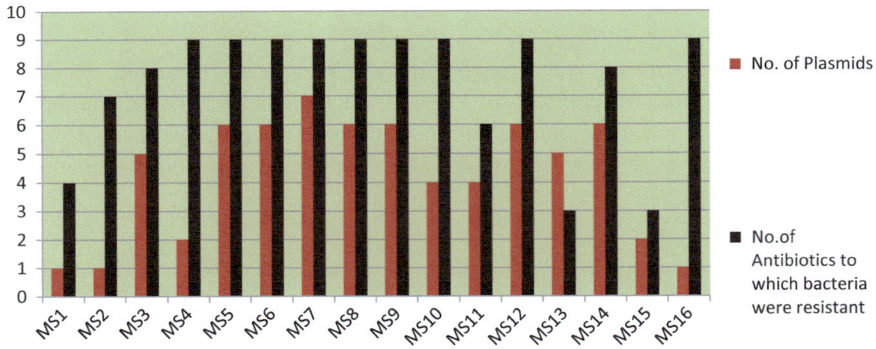

Fig. 7 Depicting the number of plasmid and number of antibiotic to which the MS+ strains were resistant

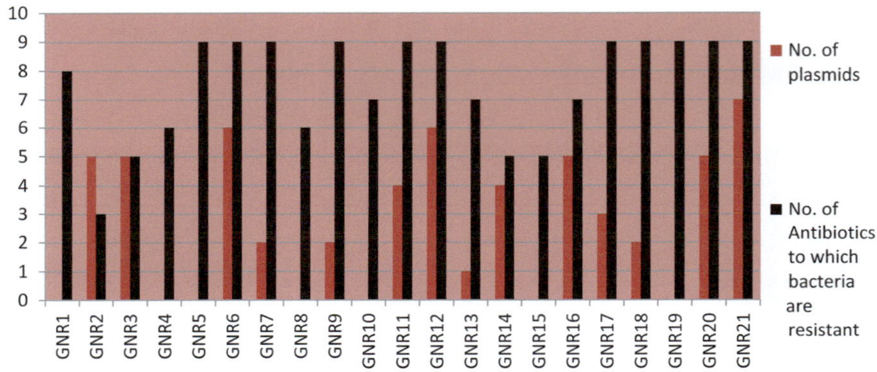

Fig. 8 Depicting the number of plasmid and number of antibiotic to which the GNR (MS−) strains were resistant

Correlation Regression Analysis

Taking two variables, i.e. (a) degree of resistance and (b) number of plasmid bands present in the bacterial strains correlation regression analyses, were made among different groups (the clinical isolates, nosocomial site strains, MS+ strains and GNR (MS−) strains) (Fig. 9).

A 'moderate positive correlation' (0.414) and a 'very weak positive correlation' (0.067) were observed in MS+ stains and GNR (MS−) strains respectively. Further,

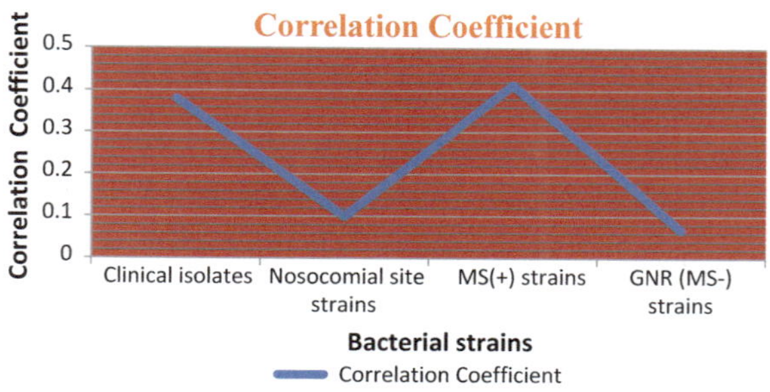

Fig. 9 Depicting the value of correlation coefficient (r) in different groups

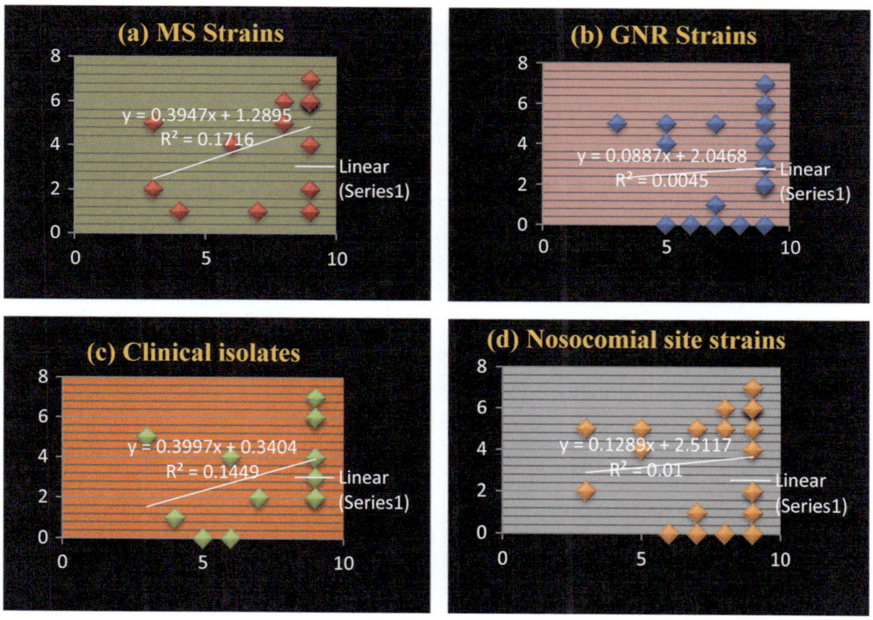

Fig. 10 **a–d** Depicting the regression plot with two variables degree of antibiotic resistance (X-axis) and number of plasmids present (Y-axis)

a weak uphill positive correlation (0.38) and very weak positive correlation (0.099) were observed with clinical and nosocomial site strains with respect to their resistance pattern and plasmid bands illustration. Finally, a positive correlation was observed between the plasmid content and the number of antibiotics to which the bacteria were resistant (Fig. 10).

Conclusion

From the statistical analysis, we can virtually conclude that, there exists a "very very weak to moderate positive correlation" between the plasmid content and degree of bacterial resistance towards test antibiotics. Hence, this study has inferred about the prevalence of the drug-resistant phenomenon among the test bacterial strains isolated from VIMSAR, Burla, Odisha, was not strongly correlated with plasmid acquisition. Therefore, the drug resistance could be due to factors present in the chromosomal DNA. However, the presence of drug-resistant genes in plasmids also cannot be denied. In this discussion, further study is recommended.

Ethical Clearance Statement

The bacterial samples used were isolated from human patients in a clinical manner. The isolation process was acknowledged by the department of Microbiology, VIMSAR, Burla, Odisha. The Institutional ethical Committee "VIREC" (Institutional Research Ethical Committee, VIMSAR), bearing Registration No. ECR/861/Inst/OR/2016 had provided the required approval for the said isolation process. The isolation process had been carried out under the approval no. 2014/P-I-RP/76, that had been provided by the institutional ethical committee. It is all the more important to mention that the isolation process was carried out from the urine samples of the patients, rather than blood samples. As would have been the case, it is necessary to have a filled-in and duly signed consent form of the concerned patient, for the collection of blood samples. But, as far as the collection of urine samples is considered, the patient's consent form is not a necessary condition.

Acknowledgements This work is financed by DST (INSPIRE), New Delhi currently being carried out in the School of Life Sciences, Sambalpur University, Odisha. This communication is a part of Ph.D. thesis work, Sambalpur University

References

Adnan SJ (2015) Antimicrobial resistance pattern and plasmid profile of some salmonella spp. Isolated from clinical samples in Riyadh area. Eur Sci J 11:136–143

Bauer AW, Kirby WMM, Sherris JC, Turck M (1966) Antibiotic susceptibility testing by a standardized single disk method. Am J Clin Pathol 36:493–496

Odeyemi AT, Awokunmi EE, Adebayo AA (2015) Plasmid profile of multi-drug resistance bacteria isolated from available water sources and leachate samples from dumpsite at Ebira communities in Ekiti north senatorial district, Ekiti state, Nigeria. Eur J Adv Res Biol Life Sci 3:32–35

Pattnaik S, Padhan M, Narayani M (2013) Resistance pattern acquired by nosocomial pathogens: a plasmid profile study. J Appl Pharm Sci 3(09):057–059

Uma B, Prabhakar K, Rajendran S, Kavitha K, Sarayu YL (2009) Antibiotic sensitivity and plasmid profiles of *Escherichia coli* isolated from Pediatric Diarrhea. J Glob Infect Dis 1(2):107–110

Optimization of Surface Sterilization Process of Selected Dye-Yielding Plants for Isolation of Bacterial Endophytes

Bushra Khanam and Ramesh Chandra

Abstract

Sterilizing agents, such as sodium hypochlorite (NaOCl), hydrogen peroxide (H_2O_2), mercuric chloride ($HgCl_2$), and ethanol (C_2H_5OH), are commonly used in plant tissue culture laboratory. The following studies were performed to estimate the effect of sterilizing agents and explants were tested for sterilization by altering concentration and time of exposure of sterilizing agents. Time duration for treatments and concentration of sterilizing agents were considered as influencing parameters. The contamination and survival percentage methods were statistically analyzed for evaluating optimum conditions for the surface sterilization of explants (root, stem, and leaf). The results revealed that as time of exposure and concentrations of sterilizing raised, the contaminations of explants declined and the combination of sterilizing agents that were mercuric chloride (0.1% for 5–10 min) and ethanol (70% for 6–8 min) showed the best result which exhibited its sterilization efficiency.

Keywords

Sterilizing agents · Dye-yielding plants · Environmental hazard

B. Khanam · R. Chandra (✉)
Department of Bio-Engineering, Birla Institute of Technology, Mesra, Ranchi 835215, Jharkhand, India
e-mail: rameshchandra@bitmesra.ac.in

K. Mukhopadhyay et al. (eds.), *Applications of Biotechnology for Sustainable Development*, DOI 10.1007/978-981-10-5538-6_7

Introduction

A sterilizing agent is defined as a chemical that kills or destroys all kind of microorganism, with the exception of bacterial spores; this refers to agents used on inanimate objects (Baddour 2008). Sterilizing agents are acting on the bacterial wall, cytoplasmic membrane (Dauphin 1988), energy metabolism (Guellouzh 1987), bacterial spores (Russell 1983), etc.; there are different types of sterilizing agents used: physical and chemical. In the case of heat-sensitive objects or explants, chemical sterilizing agents are preferred than the physical one chemical sterilizing agents includes sodium hypochlorite, hydrogen peroxide, mercuric chloride, ethanol, and many more. Different types of microbes are commonly found on the surface of plants parts like root, stem, leaf, seed, and flowers. Sterilization is one of the good techniques to sterilize the surface of explants and it is a mandatory step in plant tissue culture technique. The majority of contamination is caused by fungal, yeast, and bacterial contaminant (Leifert et al. 1994).

The term endophyte (Gr. *endon*, within; *phyton*, plant) was first coined in 1866 by De Bary. An endophyte can be defined as microorganisms such as fungi or bacteria that depend on either the complete or part of its life cycle within the healthy tissues of a living plant, typically causing no symptoms of disease (Tan and Zou 2011; Gunatilaka 2006). Endophytic bacteria are found in stem, leaves, roots, seeds, fruits, ovules, tubers, and inside legume nodules. A variation in the populations of both indigenous and introduced endophytes is due to time of sampling, environment conditions, plant source, tissue type, and their age. Generally, bacterial populations are in decreasing order from roots, stem, and leaves. In most plants, generally roots have greater numbers of endophytes compared with aboveground tissues (Rosenbluenth and Romero 2006). Bacterial endophytes have been isolated from leaves, stems, and fruits of various plant species (Kobayashi and Palumbo 2000). Isolation process is an important and critical step; it should be sensitive enough to recover endophytic microorganisms but at the same time be strong enough to eliminate epiphytes from the plant surface.

The present study reported that the influences of sterilizing agents on explants by varying their concentration and duration of exposure and their sterilization efficiency on the basis of percentage of survival of explants were described.

Materials and Methods

Collection of Plant Materials

The samples were collected from the medicinal garden of Birla Institute of Technology, Ranchi. For isolation of endophytes, healthy and mature plants were carefully chosen; Leaves, roots, and stems of selected plants were carefully cut out and collected in sterile zip-lock plastic bags.

Methods Used for Surface Sterilization

The surface sterilization is the initial and mandatory step for isolation of endophyte in order to kill all the surface microbes. It is usually accomplished by the treatment of plant tissues with sterilizing agents for a certain period and then by 3–5 times sterile rinse. Theoretically, the sterilizing agent should kill any microbes on the surface of plants material without affecting the host tissue and endophytes. In general, isolation and purification of endophytes from plant tissue consist of the following two phases. For sterilization of explants, process was divided into two: Phase I (outside Laminar Air Flow) and Phase II (inside Laminar Air Flow). Four different kinds of sterilizing agents, viz., mercuric chloride ($HgCl_2$), sodium hypochlorite (NaOCl), hydrogen peroxide (H_2O_2), and ethanol (C_2H_5OH) are tested for explants sterilization by varying their concentration and time of exposure.

Results and Discussions

Explant Sterilization

After observing the inoculated explants about one week for contamination, it was found that combination of ethanol and $HgCl_2$ significantly reduced contamination. Sodium hypochlorite (NaOCl) and hydrogen peroxide (H_2O_2) were found less effective because of being mild sterilizing agent provided more percentage of infection same in the case of ethanol too.

Effectiveness of Surface Sterilization

To check the efficiency of mercuric chloride, ethanol and sodium hypochlorite as effective sterilizing agents on survival percentage of treated explants, sterilizing agents used individually as well as in combination (Fig. 1).

The most important treatment prior to endophyte isolation from different parts of the plant is surface sterilization. Surface sterilization of plant parts (explants: root, stem, and leaf) before subjecting them for endophytes isolation is essential for the pure endophytes isolation that ensures the reduction of the contaminants, depending on their morphological characteristics like softness or hardness of the tissues. Therefore, in the present study, four sterilizing agents in different concentrations with varying time of exposure were tested for sterilization of explants of selected dye-yielding plants for isolation of bacterial endophytes. In case of *Curcuma longa* (Turmeric) 81% survival percentage was obtained with a combination of 70% ethanol for 8 min and 0.1% (w/v) $HgCl_2$ for 10 min. In *Lawsonia inermis* (Henna) 95% survival percentage was obtained with a combination of 70% ethanol for

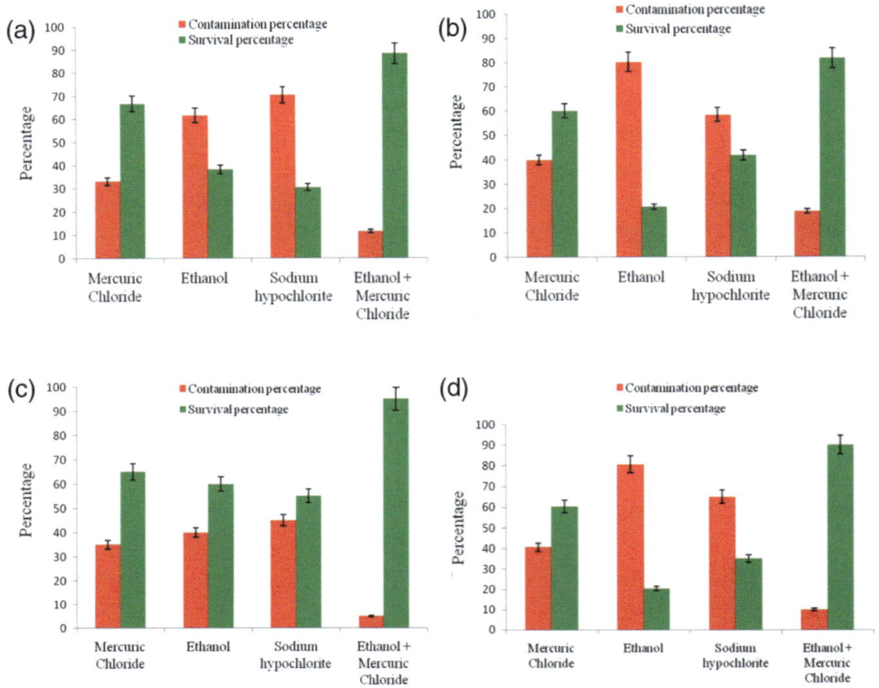

Fig. 1 The graph shows the influence of different sterilizing agents on **a** *Beta vulgaris* L. (Red Beet), **b** *Curcuma longa* L. (Turmeric), **c** *Lawsonia inermis* L. (Henna), and **d** *Zingiber officinale* L. (Ginger)

5 min and 0.1% (w/v) $HgCl_2$ for 5 min. For *Zingiber officinale* (Ginger) 90% survival percentage was obtained with a combination of 70% ethanol for 8 min and 0.1% (w/v) $HgCl_2$ for 10 min; and in *Beta vulgaris* (Beetroot) 88% survival percentage was obtained with a combination of 70% ethanol for 8 min and 0.1% (w/v) $HgCl_2$ for 10 min in all four plants. Explants showed a significant reduction in both the bacterial and fungal contamination, while other three sterilizing agents did not give acceptable results even on increased time and concentration. The result showed some similarity with other previous studies on various medicinal plants that are *Asparagus densiflorous* and *Podophyllum hexandrum* (Dasgupta et al. 2007; Sultan et al. 2006). Optimized concentration of $HgCl_2$ showed good results in the case of *Inula racemosa* (Jabeen et al. 2007) and *Picrorhiza kurroa* (Sood and Chauhan 2009). The present study disclosed that the protocols developed for the surface sterilization of *B. vulgaris* L. (Red Beet), *C. longa* L. (Turmeric), *Z. officinale* L. (Ginger), and *L. inermis* L. (Henna) have the scale-up isolation of valuable pure colonies of endophytes.

Conclusions

The thorough review of the earlier studies disclosed that there is negligible published data on surface sterilization of *B. vulgaris* L. (Red Beet), *C. longa* L. (Turmeric), *Z. officinale* L. (Ginger), and *L. inermis* L. (Henna) surface sterilization of optimized sterilization method for endophytes isolation technique. Surface sterilization of explants is an initial and vital step for isolation of endophytes; if surface sterilization was not done correctly contamination hindered to distinguish between endophytes and epiphytes. Endophytes are those microorganisms that reside in the plant tissues and are promising, less explored, and useful sources of novel natural products for exploitation in agricultural and medicinal industries.

Acknowledgements The financial support by Maulana Azad National Fellowship, in the form of SRF (F1-17.1/2013-14/MANF-2013-14-MUS-JHA-21819/(SA-III/website)) to Mrs. Bushra Khanam is acknowledged. Facilities provided by sub-DIC Bioinformatics center of Department Bio-Engineering of Birla Institute of Technology, Mesra are gratefully acknowledged.

References

Baddour MM, Abuelkheir MM, Fatani AJ (2008) A study of the effects of different disinfectants used in riyadh hospitals and their efficacy against methicillin resistant *Staphylococcus aureus* (MRSA). Saudi Pharm J 16:165–170

Dasgupta CN, Mukhopadhyay MJ, Mukhopadhyay S (2007) Somatic embryogenesis in *Asparagus densiflorous* (Kunth) Jessop cv. Sprengeri. J Plant Biochem Biotechnol 16:145–149

Dauphin A (1988) Hygiene hospitaliere practiue, 2nd edn, p 715

De Bary A (1866) Morphologie and physiologie der pilze, Flechten and *Myxomyceten.* Hofmeister's handbook of physiological botany, vol 11. Leipzig, Germany

Guellouzh H (1987). Thesis No. 304 National Veterinary School, Tunis, p 198

Gunatilaka AAL (2006) Natural product from plant associated microorganisms: distribution, structural diversity, bioactivity and implication of their occurrence. J Nat Prod 3:509–526

Jabeen N, Shawl AS, Dar AH, Jan A, Sultan P (2007) Micro propagation of *Inula racemosa* Hook. f. A high valuable medicinal plant. Int J Bot 3:296–301

Kobayashi DY, Palumbo JD (2000) Bacterial endophytes and their effects on plants and uses in agriculture. In: Bacon CW, White JF (eds) Microbial endophytes. Dekker, New York, pp 199–236

Leifert C, Morris C, Waites WM (1994) Ecology of microbial saprophytes and pathogens in field grown and tissue cultured plants. CRC Crit Rev Plant Sci 13:139–183

Rosenbluenth M, Romero EM (2006) Bacterial endophytes and their interaction with hosts. J Plant Physiol 19:327–331

Russell AD (1983) Principles of antimicrobials activity, 3rd edn. Lea & Febiger, Philadelphia, pp 717–745

Sood H, Chauhan RS (2009) Development of low cost micropropagation technology for an endangered medicinal herb *(Picrorhiza kurroa)* of North-Western Himalayas. J Plant Sci 56:1–11

Sultan P, Shawl AS, Ramteke PW, Jan A, Chisti N, Jabeen N, Shabir S (2006) *In-Vitro* propagation for mass multiplication of *Podophyllum hexandrum*: a high value medicinal herb. Asian J Plant Sci 5:179–184

Tan RX, Zou WX (2011) Endophytes: a rich source of functional metabolites. J Nat Prod 8:448–459

Molecular Biology, Genomics and Bioinformatics Insights into Fungal Pectin Lyase: An overview

S. Yadav, P.K. Yadav, A.K. Dubey, G. Anand, A. Tanveer, R. Dwivedi and D. Yadav

Abstract

Pectinase represents an industrially important group of enzymes, comprising pectin lyases (PNL), pectate lyases (PL), polygalacturonases (PG), and pectin methylesterases (PME) as a well-studied member with diverse applications. Among pectinases, pectin lyase occupies unique positions based on its reaction mechanisms. Pectin lyase is associated with degradation of pectin polymer directly by β-elimination, and 4, 5-unsaturated oligogalacturonide is the product formed. In case of other pectinases, sequential degradation of pectin molecule occurs. Further, its relevance in fruit juice clarification is mainly because of two reasons. First, they are known to degrade pectin without altering the ester group responsible for the specific aroma of the juice and second, the highly toxic methanol is not generated by the process of degradation of pectin-by-pectin lyases. The applications of pectin lyases in retting processes have recently been elucidated. There exist substantial reports of fungal pectin lyases especially from *Aspergillus* and *Penicillium* species, with sole emphasis on production optimization, purification, and biochemical characterizations. As far as molecular biology of pectin lyases is concerned, few studies have been performed especially on *Aspergillus* species and molecular cloning and expression of relevant pectin

S. Yadav · P.K. Yadav · A.K. Dubey · G. Anand · A. Tanveer · R. Dwivedi · D. Yadav (✉)
Department of Biotechnology, D.D.U Gorakhpur University, Gorakhpur 273009, UP, India
e-mail: dinesh_yad@rediffmail.com

P.K. Yadav
Department of Biological Chemistry, University of Michigan Medical School, Ann Arbor, MI 48109-0606, USA

A.K. Dubey
Department of Postharvest Science of Fresh Produce, Agricultural Research Organization, The Volcani Center, Bet Dagan 50250, Israel

K. Mukhopadhyay et al. (eds.), *Applications of Biotechnology for Sustainable Development*, DOI 10.1007/978-981-10-5538-6_8

lyases genes have been reported. Efforts have been made to decipher the factors influencing pectinases gene regulation, though not extensively. Molecular cloning of pectin lyase gene from diverse sources, its manipulation by using directed evolution techniques, and searching for novel sources by utilizing metagenomic-based approach are some of the recent developments in pectin lyase research, which is a subject matter for the present review.

Keywords
Pectin lyase · Pectinases · Bioinformatics · Molecular biology · Metagenomics · Directed evolution

Introduction

Enzymes are the integral part of human life and over the years, substantial efforts have been made to influence diverse aspects of enzyme technology like production, characterization, manipulation, and innovations (Fig. 1). Based on sources, enzymes can be plant enzymes, microbial enzymes, and animal enzymes. Microbial

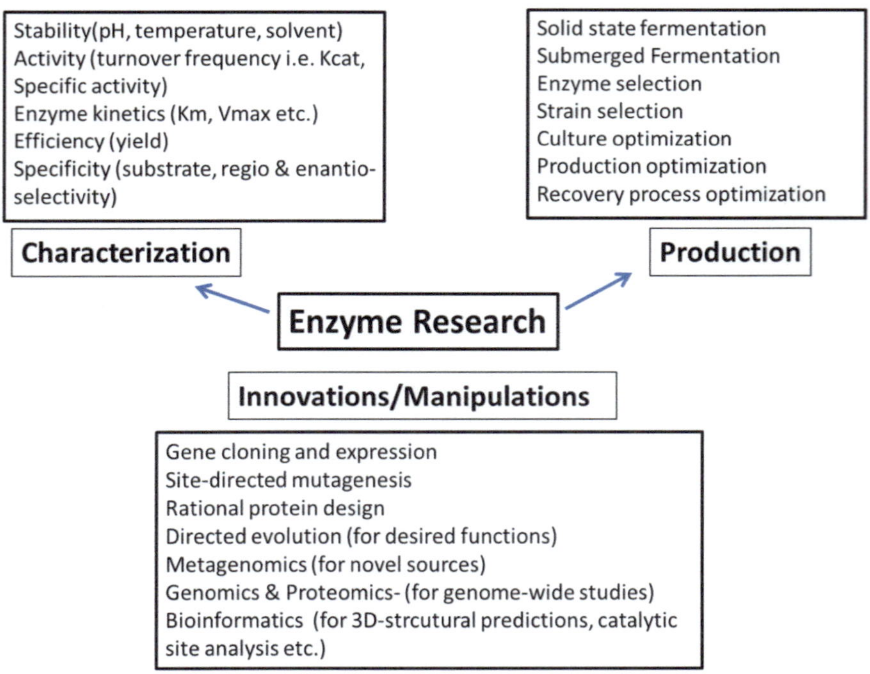

Fig. 1 Overview of trends in enzyme research

enzymes have immense potential in industries and are preferred based on stability, enzyme activity, ease of cultivation, and susceptibility to gene manipulation (Anbu et al. 2013; Binod et al. 2013; Adrio and Demain 2014). The research trend in enzyme technology has witnessed substantial changes over the years, owing to the development of state-of-the-art technologies in the form of directed evolution (Currin et al. 2015), metagenomics (Faust et al. 2015), bioinformatics, and relevant science of omics (Damborsky and Brezovsky 2014). At present, almost 4000 enzymes are known, and of these, approximately 200 enzymes used commercially are of microbial origin. However, only about 20 enzymes are produced on truly industrial scale (Li et al. 2012). In search of ideal biocatalyst, several parameters namely stability, activity, efficiency, and specificity are monitored (Lorenz and Eck 2005). The major constraint of production of microbial enzymes by conventional fermentation technology, either submerged (SmF) or solid-state fermentation (SSF), are somehow taken care by the advent of gene cloning and expression technologies in the recent years. There exist several classes of industrially important enzymes, pectinases being one of them, which is a subject matter for the present review.

Pectin Lyases

Pectinases represent a group of enzymes associated with degradation of pectin or pectic substances, known to be ubiquitously present in plant kingdom representing the major component of middle lamella of a plant cell wall, conferring firmness and structure to the plant cells and has been substantially reviewed over the years (Alkorta et al. 1998; Gummadi and Panda 2003; Lang and Dornenburg 2000; Kashyap et al. 2001; Hoondal et al. 2002; Jayani et al. 2005; Gummadi and Kumar 2005; Gummadi et al. 2007; Pedrolli et al. 2009; Payasi et al. 2009; Khan et al. 2013; Sharma et al. 2013; Kohli and Gupta 2015). Based on mode of action and substrates, pectinases have been classified into several important groups like

(i) Pectinesterases [PE, E.C. 3.1.1.11]
(ii) Polygalacturonases [PG, E.C. 3.2.1.15]
(iii) Pectate lyases [PL, E.C. 4.2.2.2]
(iv) Pectin lyases [PNL, E.C. 4.2.2.10].

The major applications of pectinases are summarized in Table 1.

Pectin lyase (EC 4.2.2.10) has a unique β-elimination reaction mechanism to degrade pectin and results in the formation of 4, 5 unsaturated oligogalacturonates. Further, the advantage of pectin lyases in fruit juice clarification industries is that it generally does not alter the ester content of the polymer chain known to be responsible for generating specific aroma of fruits. Another advantage is that it does not produce toxic methanol. The diverse sources of pectin lyases, purification strategies, enzymatic characterization, mechanism and crystallography-based structural insights, gene cloning, and expression studies along with applications

Table 1 Application of pectinases

Cloud stabilization
Fruit juice clarification
Extraction of juices and oil
Manufacture of fruit purees
Deskinning of orange segments
Maceration
Liquefaction
Gelation
Wood preservation
Retting of natural fibers
Bioscouring of cotton fibers
Processing of textile fibers
Degumming of plant bast fibers
Paper making
Pectic waste water pretreatment
Coffee and tea fermentation
Functional food
Animal feed
Purification of plant viruses
Plant disease control

have been exclusively reviewed (Yadav et al. 2009a). The application of pectin lyases from different fungal strains in retting of natural fiber has been elucidated in our lab (Yadav et al. 2008, 2009b, c, 2013, 2014, 2016, 2017). This review is an attempt to update the molecular biology, genomics, and bioinformatics aspects of pectin lyases along with other pectinases.

Molecular Cloning and Expression of Pectin Lyases

The predominant microbial source of pectin lyase reported so far are fungi though few bacteria and yeasts are also reported (Gummadi and Kumar 2005; Gummadi et al. 2007). There are few reports of cloning of pectin lyases genes, predominantly from *Aspergillus* species. Attempts have been made to understand the factors influencing the expression of pectinase genes and reports reveal that it is regulated at the transcriptional level and is influenced by environmental factors, such as pH of medium, carbon source, pectin, and pectic components (Lahiani and Gargouri 2007). In case of phytopathogenic bacteria, i.e., *Erwinia*, attempts have been made to elucidate the key elements forming the regulatory network that control the expression of pectinase genes (Rouanet et al. 1999), while regulation of fungal pectinases is still to be explored. The role of pectin lyase in disease development has been documented by a pectin lyase (pnl)-deficient mutant causing pathogenicity attenuation (Pirhonen et al. 1991). The expression of 26 pectinolytic genes from *Aspergillus* subjected to different growth conditions has been analyzed (de Vries et al. 2002). In case of *Aspergillus niger* (Gysler et al. 1990; Kusters-van Someren et al. 1991, 1992), *Aspergillus oryzae* (Kitamoto et al. 2001a) and *Glomerella cingulata* (Templeton et al. 1994), several pectin lyase genes have been amplified and the sequences obtained were characterized by specific elements. The *pelA*, *pelB*, and *pelD* genes of *A. niger* showed putative TATA box at -126, -183, and -148 positions from the transcription initiation site, respectively. Putative CAAT box is situated at -223 and -88 in *pelA* and *pelD*, respectively. The length of the coding region of these genes shows uniformity and possesses four introns, two of which are common amongst all the three genes. The coding regions of pel1 and pel2 genes of *A. oryzae* are 1196 and 1306 bp interrupted by one and three introns, respectively. In pel1 gene there is a potential TATA box at -116, CCAAT sequence at position -386, and CT-rich region of 60 bp between TATA box and ATG. In case of pel2, the putative TATA box is at position -150; however, CAAAT sequence is found at three positions, -339, -414, and -487. CT-rich region between TATA box and ATG is 85 bp long. The pectin lyase gene cloned from *G. cingulata* reveals some strikingly different features like the presence of six introns whose positions are different from intron sites found in pectin lyase genes in *A. niger*.

In case of *Penicillium griseoroseum* CCT6421, pectin lyase genes designated as *plg*1 and *plg*2 were amplified and sequence characterized. Further, attempts have been made to analyze the expression of these genes using techniques like northern hybridization and RT-PCR (Cardoso et al. 2008). The plg1 gene (pectin lyase 1) from *P. griseoroseum* under the control of strong constitutive promoter of the glyceraldehydes-3-phosphate dehydrogenase gene (*gpd*A) and terminator region of tryptophan synthase gene (*trp*C) from *Aspergillus nidulans* has been cloned and overexpressed (Cardoso et al. 2008, 2010). The cloning and expression of the *plg*1 and *plg*2 from *P. griseoroseum* have been reported earlier (Bazzolli et al. 2006). The pectin lyases genes reported from different fungal sources have been summarized in Table 2.

The studies on gene regulation and role of different regulatory elements are not substantial in case of pectinases, though some information regarding the regulatory elements upstream the promoter regions have been reported in *Aspergillus*. There are several elements like the Hap2-3-4, CreA, and PacC observed in pectinases which are considered to occur frequently in genes coding for another hydrolase, as the factors like oxygen, glucose, and the pH of the medium are common that influence the gene expression (Aro et al. 2005). In case of *Aspergillus* pectinolytic genes, a unique sequence CCCTGA associated with regulation has been observed, though the corresponding target regulatory protein could not be isolated (Benen et al. 1996; Visser et al. 2004). It has been shown that mutation of trans-regulatory transcriptional factor in *Penicillium occitanis* results in constitutive overexpression of pectin lyase, polygalacturonase, and pectate lyase on media containing different carbon sources (Ayadi et al. 2011). It shows that constitutive hyper production is controlled at the transcriptional level. Gene copy number is another important factor affecting the expression of pectinase genes. An increase in gene copy number could

Table 2 List of pectin lyase genes reported from different fungal sources

Pectin lyase genes	Accession number	Organisms	References
*Pel*A	AAA24845	*Erwinia carotovora*	Lei et al. (1988)
*Pel*B	CAA39305	*Aspergillus niger*	Harmsen et al. (1990)
*Pel*D	AAA32701	*Aspergillus niger*	Gysler et al. (1990)
*Pel*C	AAW03313	*Aspergillus niger*	Harmsen et al. (1990)
Pnl-1	BAB82467	*Aspergillus oryzae*	Kitamoto et al. (2001a, b)
Pnl-1	BAB82468		
Pnl-2	BAB82468		
Pnl-1	AF158256	*Colletotrichum gloeosporioides* f. sp. *malvae*	Wei et al. (2002)
Pnl-2	AF156984		
Pnl-2	AAD43565		
Pectin lyase F	CAD34589	*Aspergillus niger*	de Vries et al. (2002)
*Pel*A	XP_001930850	*Aspergillus terreus* NIH2624	
*Pel*B	Q5BA61	*Aspergillus nidulans*	Bauer et al. (2006)
Pectin lyase	EAL91586	*Aspergillus fumigatus Af293*	Nierman et al. (2005)
Pectin lyase	ABF50854	*Aspergillus nidulans*	Bauer et al. (2006)
Pectin lyase (*plg1*)	AAM23008	*Penicillium griseoroseum*	Bazzolli et al. (2006)
Pectin lyase (*plg2*)	AF502280		
*Pel*A	CAK48529	*Aspergillus niger*	Pel et al. (2007)
Pnl-1	ABH03046	*Penicillium occitanis*	Trigui-Lahiani et al. (2007)
*Pel*A	ABO38859	*Aspergillus nidulans*	Zhao et al. (2008)
*Pel*E	B0Y0L8	*A. fumigatus A1163*	Fedorova et al. (2008)
Afpnl-1 to Afpnl-5	JQ735890-JQ735894	*Aspergillus flavus NIICC8142*	Dubey et al. (2014)
Pectin lyase	AGO18310	*Talaromyces purpureogenus*	Perez-Fuentes et al. (2014)
Pectin lyase	AIX03726	*Aspergillus niger*	Xu et al. (2014)
Afpnl-1, *Afpnl*-2, *Afpnl*-3, *Afpnl*-4, *Afpnl*-5, *Afpnl*-6, *Afpnl*-7	AFLA_119860 AFLA_124660 AFLA_116040 AFLA_025400 AFLA_017180 AFLA_14060 AFLA_007720	*Aspergillus flavus* NRRL 3357	Dubey et al. (2016a)
Pectin lyase	KJ729121	*Aspergillus niger* MTCC 404	Dubey et al. (2016b)

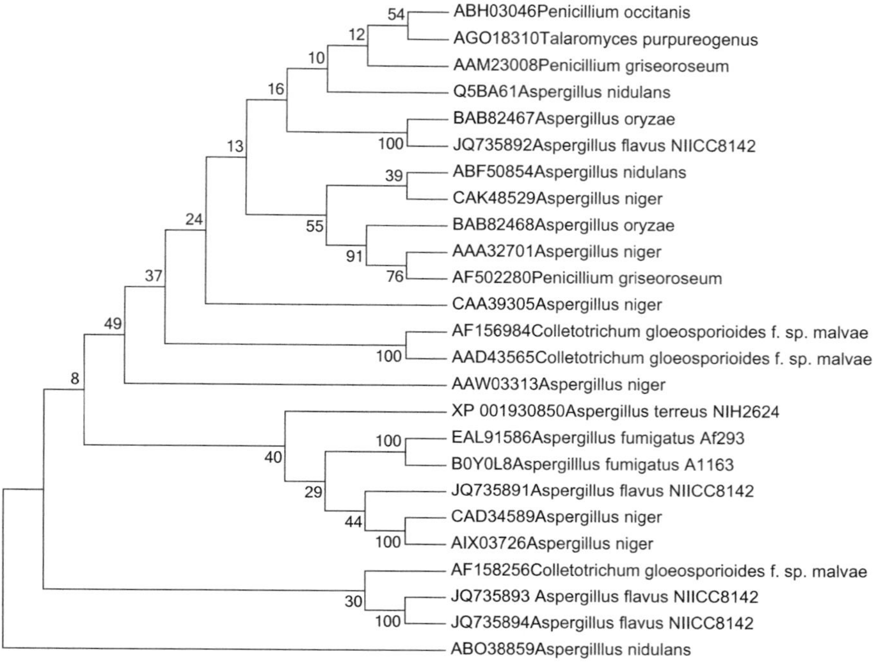

Fig. 2 Phylogenetic tree showing pectin lyase genes cloned from different fungal sources by NJ method

increase the enzyme production; however, transcription is affected by gene integration position (Teixeira et al. 2014). Attempts have been made to analyze the expression of several recombinant pectin lyase genes from a mutant strain of *Penicillium citrinum* subjected to different carbon sources. The recombinant pectin lyase gene revealed the highest activity in response to sugar cane juice (Teixeira et al. 2011).

A total of 18 pectin lyase genes amplified from different species of *Aspergillus* namely *A. niger*, *A. versicolor*, *A. foetidus*, *Aspergillus terricola*, *A. aculeatus*, *A. ficcum*, *Aspergillus flavus*, and *P. citrinum* were sequenced. The assigned accession numbers of these pectin lyases genes were from JF447757 to JF447774. The bioinformatics-based assessment of these pectin lyase genes for deciphering the relatedness by phylogenetic tree construction (Fig. 2), motif assessment, and multiple sequence alignment has been attempted (Dubey et al. 2012a).

A total of five pectin lyase genes from *A. flavus* NIICC8142 with GenBank accession numbers JQ735890, JQ735891, JQ735892, JQ735893, and JQ735894 have been characterized using bioinformatics tools (Dubey et al. 2014). The BLAST search revealed significant homology with the reported pectin lyase genes, and multiple sequence alignment among these pectin lyases showed several stretches of conserved sequences. The uniform presence of pec_lyase C domain was also observed among these sequences (Dubey et al. 2014). Molecular cloning and bioinformatics assessment of pectin lyase gene from *Aspergillus niger* MTCC 404 has been recently reported (Dubey et al. 2016b).

Bioinformatics of Pectin Lyases

The web-based resources and tools of Bioinformatics are significantly influencing the direction of enzyme research. The assessment of nucleotide or protein sequences available in databases is being investigated for deciphering the percentage homology by subjecting to BLAST search, revealing stretches of conserved regions among several sequences by Multiple Sequence Alignment (MSA) tools, reducing the degree of relatedness by constructing phylogenetic trees by different methods like NJ method, Maximum Likelihood method, motif assessment, bioinformatics-based 3-D structural prediction, revealing gene organization, determining physiochemical attributes, etc. The protein sequences of pectin lyases along with other pectinases, representing different microbial sources, were computationally analyzed and revealed specific motifs associated with the typical mechanism of pectin lyases (Yadav et al. 2009d). Similar to pectin lyases, the available protein sequences of other pectinases namely pectate lyases and polygalacturonases, representing different source organisms, were also in silico characterized (Dubey et al. 2010, 2012b). The analysis of 18 pectin lyase genes representing different species of *Aspergillus* and *P. citrinum* and five pectin lyase genes of *A. flavus* NIICC8142 has been attempted using different bioinformatics tools (Dubey et al. 2012a, 2014). A cloned pectin lyase gene from *Colletotrichum lindemuthianum* has been extensively subjected to bioinformatics analysis, especially comparative phylogenetic and elucidation of structures based on homology modeling, to reveal the relationship between the pectin lyase genes with genes from phytopathogenic and saprophytic/opportunistic microorganisms (Lara-Marquez et al. 2011). In silico characterization of genes encoding cell wall degrading enzymes in *Phytophthora parasitica* revealed the presence of 44 polysaccharide lyases (PL) which occurred in PL1, PL3, and PL4 families. PL1 contains pectin lyase that can catalyze esterified α-1, 4-galacturonan residues at the non-reducing end of homogalacturonan (Blackman et al. 2014). The availability of genome sequence of model dicot plant *Arabidopsis thaliana* led to studies related with genome-wide distribution of pectin lyases. An extensive in silico characterization of the identified pectin lyase genes from the genome of *A. thaliana* along with real-time expression analysis has been reported. An insight to several attributes like phylogenetic relatedness, position of the genes on the chromosomes, possibility of gene duplication events, intron–exon organization, rate of functional divergence, and evolutionary aspects have been emphasized in this study (Cao 2012). Homology-based three-dimensional structural predictions, validation and analysis of 28 pectate lyases, representing different species of *Aspergillus* namely *A. flavus, A. niger, A. nidulans, Aspergillus fumigatus, A. oryzae, A. clavatus, and Aspergillus terreus* revealed structural diversity (Kumar et al. 2014). An insight into three dimensional structures of pectin lyases of *Aspergillus flavus* NRRl 3357 has been elucidated using in-silico tools (Dubey et al. 2016a).

Directed Evolution of Pectin Lyases

Directed evolutions, the state-of-the-art technology, generally mimics Darwinian natural evolutions in vitro. It comprises iterative cycles of mutagenesis, recombination, and selection or screening for the enzymes with desired features for industrial applications. The directed evolution techniques and its diverse application for manipulating enzymes for a desired feature, relevant to industrial applications, have been extensively reviewed over the years (Arnold et al. 2001; Tracewell and Arnold 2009; Bottcher and Bornscheuer 2010; Dalby 2011; Kumar and Singh 2012; Denard et al. 2015). In case of pectinases group of enzymes, using site-directed mutagenesis, amino acid residues critical for catalysis and stability have been analyzed for *A. niger* family 1 pectin lyase A (Sanchez-Torres et al. 2003). An insight into the unique β–elimination mechanism of pectin lyase was assessed by such study, which cleaves the α-D-(1-4) glycosidic bond between galacturonosyl residues in homogalacturonan part (smooth) of the pectin molecule to form a double bond at the non-reducing end of the cleaved polysaccharides. The directed evolution technique has been attempted for pectate lyases, showing efficient bioscouring of cotton fabric. A substantially higher optimum temperature of the evolved enzyme obtained seems to be highly effective as compared to chemical scouring method (Solbak et al. 2005). A pectin methylesterase A from *Erwinia chrysanthemi* has been subjected to direct evolution for enhanced thermal stabilization, making it more suitable for applications in sugar beet pulp biorefinery (Chakiath et al. 2009).

Metagenomics Approach in Search of Novel Sources

Substantial efforts have been made to screen indigenous fungal strains showing appreciable production of pectin lyases and further optimized either by solid-state fermentation or submerged fermentations. It is well-known fact that only 1% of the total microbial population is culturable under laboratory conditions, while 99% still remains to be explored, which may be a potential source of novel microbial enzymes. There are several reports of novel pectinases from the metagenomic-derived samples. A thermostable pectinase has been isolated and characterized from the soil metagenome showing a broad range of catalytic activity. Pectate lyase, another member of the pectinase family, is also an economically important enzyme (Fang et al. 2011; Mukhopadhyay et al. 2013). The high production cost, poor activity, and stability are major constraints for possible industrial applications. A catalytically efficient alkaline pectate lyase has been isolated from alkaline environmental samples using metagenomic approach (Wang et al. 2014). A metagenome-derived sample has also been found to be a source of polygalacturonase, a pectinase belonging to the Glycosyl hydrolase 28 (GH28) of pectinases. This enzyme derived from the diversity-rich Western Ghats of India has been subjected to molecular modeling and enzymatic characterization. Pectin lyase has also been obtained from rumen of Indian buffalo (Singh et al. 2014). Using integrated metagenomics and metaproteomics approach the microorganisms and enzymes present in the solid-state fermentation of Pu-erh tea have been analyzed, where pectinases including pectin lyase responsible for maceration and soft rotting of tea leaves have been reported (Zhao et al. 2015). Recently efforts have been made to assess different methods for metagenomic DNA isolation from different soils in search of novel pectin lyase sources (Tanveer et al. 2016).

Conclusion

The studies representing fungal pectin lyases with sole emphasis on isolation, purification, and biochemical characterization along with elucidating possible industrial applications especially in fruit juice clarifications and fiber retting have been reported. The molecular cloning and expression of fungal pectin lyases genes are comparatively less studied. The bioinformatics-based assessment of pectin lyases is also being investigated, though the 3-D structural predictions along with catalytic site determination are still in early phases. Relatively low k_{cat}/K_m value of the reported pectin lyases provides an opportunity to utilize the recent innovations to improve the catalytic efficiency. Based on the potential industrial application of the pectin lyase, there is also a need to develop appropriate strategies for enhanced production of the enzyme by cloning and overexpression of relevant pectin lyase genes. Enhancing thermostability, pH stability, and wider pH range for fungal pectin lyases, as per the need of industries, could be targeted using directed evolution technique. Further, there is a need to search for novel sources for pectin lyases using metagenomics approach. The development of various computational tools for structural predictions can give an insight into the protein structure and functions, prior to subjecting it to expensive experimental methods like X-ray crystallography and NMR. The science of 'omics' namely genomics, proteomics, and bioinformatics have tremendous potential to be applied in pectin lyase research.

Acknowledgements S. Yadav sincerely acknowledges the financial grants received by Department of Science and Technology, Government of India, New Delhi, in the form of Women Scientist Fellowship (SR/WOSA/LS-34/2004), Fast Track Young Scientist Fellowship (SR/FT-LS-125/2008), and SERB Young Scientist Fellowship (SB/FT/LS-430/2012). D. Yadav acknowledges the financial support of UGC, India, in the form of UGC-Major project (F. No. 37–133/2009-SR). UGC-Dr. D.S. Kothari Post-Doctoral fellowship to A Tanveer and JNM Scholarship for Doctoral Studies to G. Anand is also sincerely acknowledged. The authors acknowledge the infrastructural support from the Head, Department of Biotechnology, D.D.U. Gorakhpur University, Gorakhpur.

References

Adrio JL, Demain AL (2014) Microbial enzymes: tools for biotechnological processes. Biomolecules 4:117–139

Alkorta I, Garbisu C, Llama MJ, Serra JL (1998) Industrial application of pectic enzymes: a review. Process Biochem 33(1):21–28

Anbu P, Gopinath SCB, Cihan AC, Chaulagain BP (2013) Microbial enzymes and their application in industries and medicine. Bio Med Res Int 2013(204014), 2 p

Arnold FH, Wintrode PL, Miyazaki K, Gershenson A (2001) How enzymes adapt: lessons from directed evolution. Trends Biochem Sci 26(2):100–106

Aro N, Pakula T, Penttila M (2005) Transcriptional regulation of plant cell wall degradation by filamentous fungi. FEMS Microbiol Rev 29:719–739

Ayadi M, Trigui S, Trigui-Lahiani H, Hadj-Taïeb N, Jaoua M, Gargouri A (2011) Constitutive over-expression of pectinases in *Penicillium occitanis* CT1 mutant is transcriptionally regulated. Biotechnol Lett 33:1139–1144

Bauer S, Vasu P, Persson S, Mort AJ, Somerville CR (2006) Development and application of a suite of polysaccharide-degrading enzymes for analyzing plant cell walls. Proc Natl Acad Sci USA 103(30):11417–11422

Bazzolli DS, Ribon AO, de Queiroz MV, de Araújo EF (2006) Molecular characterization and expression profile of pectin-lyase-encoding genes from *Penicillium griseoroseum*. Can J Microbiol 52(11):1070–1077

Benen J, Parenicova L, Kusters-van Someren M, Kester H, Visser J (1996) Molecular genetics and biochemical aspects of pectin degradation in *Aspergillus*. In: Pectins and Pectinases, vol 14. Elsevier Science, pp 331–346

Binod P, Palkhiwala P, Gaikaiwari R, Nampoothiri KM, Duggal A, Dey K, Pandey A (2013) Industrial enzymes, present status and future perspectives for India. J Sci Ind Res 72:271–286

Blackman LM, Cullerne DP, Hardham AR (2014) Bioinformatics characterization of genes encoding cell wall, degrading enzymes in the *Phytophthora parasitica* genome. BMC Genomics 15:785

Bottcher D, Bornscheuer UT (2010) Protein engineering of microbial enzymes. Curr Opin Microbiol 13:274–282

Cao J (2012) The pectin lyases in *Arabidopsis thaliana*: evolution, selection and expression profiles. PLoS ONE 7(10):e46944

Cardoso PG, Ribeiro JB, Teixeira JA, de Queiroz MV, de Araújo EF (2008) Overexpression of the plg1 gene encoding pectin lyase in *Penicillium griseoroseum*. J Ind Microbiol Biotechnol 35 (3):159–166

Cardoso PG, Teixeira JA, de Queiroz MV, de Araújo EF (2010) Pectin lyase production by recombinant *Penicillium griseoroseum* strain 105. Can J Microbiol 56(10):831–837

Chakiath C, Lyons MJ, Kozak RE, Laufer CS (2009) Thermal stabilization of *Erwinia chrysanthemi* Pectin Methylesterase A for application in a sugar beet pulp biorefinery. Appl Environ Microbiol 75(23):7343–7349

Currin A, Swainston N, Day PJ, Kell DB (2015) Synthetic biology for directed evolution of protein biocatalysts: navigating sequence space intelligently. Chem Soc Rev 44:1172–1239

Dalby PA (2011) Strategy and success for the directed evolution of enzymes. Curr Opin Struct Biol 21:473–480

Damborsky J, Brezovsky J (2014) Computational tools for designing and engineering enzyme. Curr Opin Chem Biol 19:8–16

de Vries RP, Jansen J, Aguilar G, Parenicova L, Joosten V, Wulfert F, Benen JA, Visser J (2002) Expression profiling of pectinolytic genes from *Aspergillus niger*. FEBS Lett 530(1–3):41–47

Denard CA, Ren H, Zhao H (2015) Improving and repurposing biocatalyst via directed evolution. Curr Opin Chem Biol 25:55–64

Dubey AK, Yadav S, Kumar M, Singh VK, Sarangi BK, Yadav D (2010) *In-silico* characterization of pectate lyases protein sequences from different source organisms. Enzyme Res 2010 (950230):11 p

Dubey AK, Yadav S, Anand G, Bisht NC, Yadav D (2012a) Insights to sequences of PCR amplified pectin lyase genes from different fungal strains. Online J Bioinf 13:80–92
Dubey AK, Yadav S, Rajput R, Anand G, Yadav D (2012b) In-silico characterization of bacterial, fungal and plant polygalacturonase protein sequences. Online J Bioinf 13(2):246–259
Dubey AK, Yadav S, Anand G, Yadav D (2014) PCR amplification, sequencing and *in-silico* characterization of pectin lyase gene from *Aspergillus flavus* NIICC8412. In: Kharwar RN, Upadhyay RS, Dubey NK, Raghuwanshi R (eds) Microbial diversity and biotechnology in food security. Springer, pp 413–421
Dubey AK, Yadav S, Tanveer A, Anand G, Singh VK, Yadav D (2016a) In silico 3D structure ofpectin lyase proteins of *Aspergillus flavus* NRRL 3357. Online J Bioinf 17(1):13–28
Dubey AK, Yadav S, Anand G, Singh VK, Yadav D (2016b) Molecular cloning in-silico analysis of pectin lyase from *Aspergillus niger* MTCC 404. Online J Bioinf 17(2):136–147
Fang S, Li J, Liu L, Du G, Chen J (2011) Overproduction of alkaline polygalacturonate lyase in recombinant *Escherichia coli* by a two-stage glycerol feeding approach. Biores Technol 102 (22):10671–10678
Faust K, Lahti L, Gonze D, de vos WM, Raes J (2015) Matagenomics meet time series analyses: unraveling microbial community dynamics. Curr Opin Microbiol 25:56–66
Fedorova ND, Khaldi N, Joardar VS et al (2008) Genomic islands in the pathogenic filamentous fungus *Aspergillus fumigates*. PLoS Genet 4(4):46 (Gene 388(1–2):54–60)
Gummadi SN, Panda T (2003) Purification and Biochemical properties of microbial pectinases-a review. Process Biochem 38(7):987–996
Gummadi SN, Kumar DS (2005) Microbial *Pectic transeliminases*. Biotech Lett 25:451–458
Gummadi SN, Manoj N, Kumar SD (2007) Structural and biochemical properties of Pectinases. In: Industrial enzymes: structure, function and application. Springer, pp 99–115
Gysler C, Harmsen JAM, Kester HCM, Visser J, Helm J (1990) Isolation and structure of the pectin lyase D-encoding gene from *Aspergillus niger*. Gene 89:101–108
Harmsen JA, Kusters-van Someren MA, Visser J (1990) Cloning and expression of a second *Aspergillus niger* pectin lyase gene (pelA): indications of a pectin lyase gene family in *A. niger*. Curr Genet 18(2):161–166
Hoondal GS, Tiwari RP, Tewari R, Dahiya N, Beg QK (2002) Microbial alkaline pectinases and their industrial applications: a review. Appl Microbiol Biotechnol 59(409):418
Jayani RS, Saxena S, Gupta R (2005) Microbial pectinolytic enzymes: a review. Process Biochem 40:2931–2944
Kashyap DR, Vohra PK, Chopra S, Tewari R (2001) Applications of pectinases in the commercial sector: a review. Biores Technol 77:215–227
Khan M, Nakkeeran E, Umesh-Kumar S (2013) Potential application of pectinase in developing functional foods. Ann Rev Food Sci Technol 4:21–34
Kitamoto N, Yosltino-Yasuda S, Ohmiya K, Tsukagoshi N (2001a) A second pectin lyase gene *(pel 2)* from *Aspergillus oryzae* KBN616: its sequence analysis and over expression, and characterization of the gene products. J Biosci Bioeng 91:378–381
Kitamoto N, Yosltino-Yasuda S, Ohmiya K, Tsukagoshi N (2001b) Sequence analysis and overexpression of a pectin lyase gene *(pel 1)* from *Aspergillus oryzae* KBN616. Biosci Biotechnol Biochem 65:209–212
Kohli P, Gupta R (2015) Alkaline pectinases: a review. Biocatal Agric Biotechnol 4:279–285
Kumar A, Singh S (2012) Directed evolution: tailoring biocatalysts for industrial applications. Crit Rev Biotechnol 1–14
Kumar M, Yadav S, Nasim J, Yadav D (2014) Computational assessment of predicted three dimensional structures of pectate lyases from different species of Aspergillus using homology modeling. Eur J Biotechnol Biosci 2(6):59–71
Kusters-van Someren MA, Harmsen JAM, Kester HCM, Visser J (1991) Structure of the *Aspergillus niger pel*A gene and its expression in *Aspergillus niger* and *Aspergillus nidulans*. Curr Genet 20:293–299

Kusters-van Someren M, Flipphi M, de Graa LH, van den Broeck H, Kester H, Hinnen A, Visser J (1992) Characterization of the *Aspergillus nigerpel* B gene: structure and regulation of expression. Mol Gen Genet 234:113–120

Lahiani HT, Gargouri A (2007) Cloning, genomic organization and mRNA expression of a pectin lyase gene from a mutant strain of *Penicillium occitanis*. Gene 388:54–60

Lang C, Dornenburg H (2000) Perspectives in the biological function and technological application of polygalacturonases. Appl Microbiol Biotechnol 53(4):366–375

Lara-Márquez A, Zavala-Páramo MG, López-Romero E, Calderón-Cortés N, López-Gómez R, Conejo-Saucedo U, Cano-Camacho H (2011) Cloning and characterization of a pectin lyase gene from *Colletotrichum lindemuthianum* and comparative phylogenetic/structural analyses with genes from phytopathogenic and saprophytic/opportunistic microorganisms. BMC Microbiol 11:260

Lei SP, Lin HC, Wang SS, Wilcox G (1988) Characterization of *Erwinia carotovora pel*B gene and its product pectate lyase A. Gene 62:159–164

Li S, Yang X, Yang S, Zhu M, Wang X (2012) Technology prospecting on enzymes: application, marketing and engineering. Comput Struct Biotechnol J 2(3):e201209017

Lorenz P, Eck J (2005) Metagenomics and industrial applications. Nat Rev Microbiol 3:510–516

Mukhopadhyay A, Dutta N, Chattopadhyay D, Chakrabarti K (2013) Degumming of ramie fiber and the production of reducing sugars from waste peels using nanoparticle supplemented pectate lyase. Biores Technol 137:202–208

Nierman WC, Pain A, Anderson MJ et al (2005) Genomic sequence of the pathogenic and allergenic filamentous fungus *Aspergillus fumigates*. Nature 438(7071):1151–1156

Payasi A, Sanwal R, Sanwal GC (2009) Microbial pectate lyases: characterization and enzymological properties. World J Microbiol Biotechnol 25:1–14

Pedrolli DB, Monteiro AC, Gomes E, Carmona EC (2009) Pectin and pectinases: production, characterization and industrial application of microbial pectinolytic enzyme. Open Biotechnol J 3:9–18

Pel HJ, de Winde JH, Archer DB et al (2007) Genome sequencing and analysis of the versatile cell factory *Aspergillus niger* CBS 513.88. Nat Biotechnol 25(2):221–231

Perez-Fuentes C, Cristina Ravanal M, Eyzaguirre J (2014) Heterologous expression of a *Penicillium purpurogenum* pectin lyase in *Pichia pastoris* and its characterization. Fungal Biol 118(5–6):507–515

Pirhonen M, Saarilahti H, Karlsson MB, Palva ET (1991) Identification of pathogenity determinates of *Erwinia carotovora* subsp. *carotovora* by transposon mutagenesis. Mol Plant-Microbe Interact 4:276–283

Rouanet C, Nomura K, Tsuyumu S, Nasser W (1999) Regulation of pelD and pelE, encoding major alkaline pectate lyases in *Erwinia chrysanthemi*: involvement of the main transcriptional factors. J Bacteriol 184:5948–5957

Sanchez-Torres P, Visser J, Benen JA (2003) identification of amino acid residues critical for catalysis and stability in *Aspergillus niger* family 1 pectin lyase A. Biochem J 370:331–337

Sharma N, Rathore M, Sharma M (2013) Microbial pectinase: sources, characterization and applications. Rev Environ Sci Bio/Technol 12(1):45–60

Singh KM, Reddy B, Patel D, Patel AK, Parmar N, Patel A, Patel JB, Joshi CG (2014) High potential source for biomass degradation enzyme discovery and environmental aspects revealed through metagenomics of Indian buffalo rumen. BioMed Res Int 2014(267189):10 p

Solbak et al (2005) Discovery of pectin-degrading enzyme and directed evolution of a novel pectate lyase for processing cotton fabrics. J Biol Chem 280(10):9431–9438

Tanveer A, Yadav S, Yadav D (2016) Comparative assessment of methods for metagenomic DNA isolation from soils of different crop growing fields. 3 Biotech (6):220

Teixeira JA, Gonçalves DB, de Queiroz MV, de Araújo EF (2011) Improved pectinase production in *Penicillium griseoroseum* recombinant strains. J Appl Microbiol 111(4):818–825

Teixeira JA, Nogueira GB, Queiroz MV, Araújo EF (2014) Genome organization and assessment of high copy number and increased expression of pectinolytic genes from *Penicillium griseoroseum*: a potential heterologous system for protein production. J Ind Microbiol Biotechnol 41:1571–1580

Templeton MD, Sharrock KR, Bowen JK, Crowhust RN, Rikkerink EHA (1994) The pectin lyase-encoding gene (pnl) family from *Glomerella cingulata:* characterization of *pnlA* and its expression in yeast. Gene 142:141–146

Tracewell CA, Arnold FH (2009) Directed enzyme evolution: climbing fitness peaks one amino acid at a time. Curr Opin Chem Biol 13(1):3–9

Trigui-Lahiani H, Gargouri A (2007) Cloning, genomic organization and mRNA expression of a pectin lyase gene from a mutant strain of *Penicillium occitanis*

Visser J, Bussink HJ, Witteveen C (2004) Gene expression in filamentous fungi. In: Smith A (ed) Gene expression in recombinant microorganisms. Marcel Dekker, Inc., New York, pp 241–308

Wang H, Li X, Ma Y, Song J (2014) Characterization and high-level expression of a metagenome-derived alkaline pectate lyase in recombinant *Escherichia coli*. Process Biochem 49(1):69–76

Wei Y, Shih J, Li J, Goodwin PH (2002) Two pectin lyase genes, pnl-1 and pnl-2, from *Colletotrichum gloeosporioides f.* sp. *malvae* differ in a cellulose-binding domain and in their expression during infection of Malvapusilla. Microbiology 148(7):2149–2157

Xu S, Qin X, Liu B, Zhang D, Zhang W, Wu K, Zhang Y (2014) An acidic pectin lyase from *Aspergillus niger* with favorable efficiency in fruit juice clarification. Lett Appl Microbiol 60 (2):181–187

Yadav S, Yadav PK, Yadav D, Yadav KDS (2008) Purification and characterization of an alkaline pectin lyase from *Aspergillus flavus*. Process Biochem 43:547–552

Yadav S, Yadav PK, Yadav D, Yadav KDS (2009a) Pectin lyase: a review. Process Biochem 44:1–10

Yadav S, Yadav PK, Yadav D, Yadav KDS (2009b) Purification and characterization of pectin lyase produced by *Aspergillus terricola* and its application in retting of natural fibers. Appl Biochem Biotechnol 159:270–283

Yadav S, Yadav PK, Yadav D, Yadav KDS (2009c) Purification and characterization of pectin lyase secreted by *Penicillium citrinum*. Biochemistry 74:800–806

Yadav PK, Singh VK, Yadav S, Yadav KDS, Yadav D (2009d) *In-silico* analysis of pectin lyase and pectinase sequences. Biochemistry 74:1049–1055

Yadav S, Dubey AK, Anand G, Yadav D (2013) Purification and characterization of pectin lyase secreted by *Aspergillus flavus* MTCC 10938. Appl Biochem Microbiol 49(4):400–405

Yadav S, Dubey AK, Anand G, Kumar R, Yadav D (2014) Purification and biochemical characterization of an alkaline pectin lyase from *Fusarium decemcellulare* MTCC 2079 suitable for *Crotalaria juncea* fibre retting. J Basic Microbiol 54(Suppl 1):S161–S169

Yadav D, Yadav S, Dwivedi R, Anand G, Yadav PK (2016) Potential of microbial enzymes in retting of natural fibers: A review. Curr Biochem Eng 3:89–99

Yadav D, Maurya SK, Anand G, Dwivedi R, Yadav D (2017) Purification, characterization and retting of *Crotolarea juncea* fibre by an alkaline pectin lyase from *Fusarium oxysporum* MTCC 1755. 3 Biotech 7(2):136

Zhao Q, Yuan S, Wang X, Zhang Y, Zhu H, Lu C (2008) Restoration of mature etiolated cucumber hypocotyl cell wall susceptibility to expansion by pretreatment with fungal pectinases and EGTA in vitro. Plant Physiol 147(4):1874–1885

Zhao M, Zhang D, Su X, Duan S, Wan J, Yuan W, Liu B, Ma Y, Pan Y (2015) An integrated metagenomics/metaproteomics investigation of the microbial communities and enzymes in solid-state fermentation of Pu-erh tea. Sci Rep 5:10117

Control of Aflatoxin Biosynthesis in Peanut with Geocarposphere Bacteria: A Biotechnological Approach for Sustainable Development

H.K. Chourasia and Prakash Kumar Sah

Abstract

Roots and pods of field-grown peanut were sampled at three developmental stages and a total of seven bacterial strains and one toxigenic *Aspergillus flavus* isolates were isolated from the geocarposphere. The biocontrol potential of each bacterial strain was tested against growth and Afl-B_1 production potential of toxigenic *A. flavus*. In greenhouse experiments, co-inoculation with toxigenic *A. flavus* and geocarposphere bacteria in root regions of 1- to 2-week-old peanut plants resulted in the lower synthesis of Afl-B_1 in the peanut kernels at maturity. Of the seven bacterial strains tested, four strains showed a reduction in aflatoxins production in varying extents. Pre-inoculation of bacterial strains (1 day earlier) resulted in greater inhibition of aflatoxin accumulation than those bacterial strains introduced 1 day after inoculation of toxigenic *A. flavus* strain. *Bacillus megaterium* showed maximum inhibition of aflatoxin biosynthesis, as compared to remaining three potential bacterial strains.

Keywords

Control · Aflatoxin · Peanut · Geocarposphere · Bacteria

H.K. Chourasia (✉)
TNB College, TM Bhagalpur University, Bhagalpur, Bihar, India
e-mail: hkchourasia96@gmail.com

P.K. Sah
Department of Home Science—Food and Nutrition, TM Bhagalpur University, Bhagalpur, Bihar, India

K. Mukhopadhyay et al. (eds.), *Applications of Biotechnology for Sustainable Development*, DOI 10.1007/978-981-10-5538-6_9

Introduction

Aflatoxins, the toxic metabolites of *Aspergillus flavus* Link and *A. parasiticus Speare*, pose serious health hazards as potent carcinogens to humans and domestic animals, because of their frequent occurrence in agricultural commodities including cereals and spices (Chourasia 2001; Chourasia et al. 2008; Mandal et al. 2010). Soil surrounding the peanut pod, i.e. the geocarposphere, contains higher populations of microorganisms than buffer soil (Kloepper and Bowen 1991).

In recent past, a wide range of chemicals and bioagents have been tested against pre- and post-harvest aflatoxin contaminations of peanut kernels (Bowen and Backman 1989). However, paucity of potential bio-competitive agents is striking particularly among geocarposphere bacteria (Dorner et al. 1990; Chourasia 1995). It was hypothesized that geocarposphere bacteria would be ideal candidates for protecting the developing peanut pods against aflatoxigenic fungi. The present study was aimed to study the biocontrol potential of geocarposphere bacteria against aflatoxin biosynthesis in peanut.

Materials and Methods

Sampling Procedures

Three samples of different stages, i.e. emergence of peg, swelling of peg and full-size pod with visible seed cotyledon, were used. Six replicate plants were randomly selected on each sampling date. Six replicate soil samples were also collected from nonplanted buffer zones and shaken in 10 ml 0.2 M phosphate buffer pH 7.0 (PB). Peg and pod samples were macerated with sterile mortars and pestles, and all samples were plated on tryptic soy agar for bacterial collection. Plates for bacteria were incubated at 28 °C for 24–96 h, and colonies were enumerated using a laser colony counter with bacterial enumeration software (Spiral System Instruments, Bethesda, Maryland). Pure culture of fungi was incubated for 7 days at 25 °C and *A. flavus* was identified under a Nikon stereomicroscope by following the manual of Barnett and Hunter (1972). Bacterial strains were identified with the help of bacterial enumeration software system and physiological characterization. The identification of bacterial strains and *A. flavus* was made by sending the pure culture to IMTECH, Chandigarh as reference.

Inoculum Preparation

A. flavus was grown on PDA and stored at 4 °C. Spores were harvested by flushing 5-day-old cultures in a sterile 0.01% solution of Triton X-100. The spore concentration was standardized to 3×10^6 spores/ml by using an improved Neubauer hemocytometer with suitable dilutions. All bacteria under study were grown on nutrient agar slopes at 28 °C and the suspension of 10^5 cells/ml was used as the inoculum potential.

Antibiosis Test of Bacteria Against *A. flavus*

Organisms were inoculated individually or in pairs on PAF (Pseudomonas Agar Fluorescent) and TSA (Tryptic Soya Agar) culture media separately. In case of dual cultures, colony diameters and the gap between the colonies were measured daily from 2 to 15 days and microbial interactions on culture media were classified by the method of Johnson and Curl (1972).

Screening of Bacteria in the Assays

Seed Assay

Surface-disinfested and dried peanut seeds (5 subsamples per bacterial strain) were soaked in each bacterial suspension for 5 min and then placed on water agar. Controls consisted of seeds soaked in sterile water. After incubation for 24 h at 28 °C, seeds were inoculated with 20 ml of a log 6 conidial suspension of *A. flavus* and incubated at 30 °C. Control samples treated with and without *A. flavus* were included. Data on fungal growth were collected.

Root-Radicle Assay

Bacterial suspensions were prepared as described above, and surface-disinfested and dried seeds with emerged radicles were soaked in each suspension for 5 min. After 24 h incubation at 28 °C, 20 ml of a log 6 conidial suspension of *A. flavus* was inoculated on the radicles 1.0 cm from the seed. Inoculated seeds were incubated and data on fungal growth, number of root branches, final root length, and time to complete radicle coverage with mycelial growth were recorded.

Inoculation of Bacteria/Toxigenic *A. flavus* in the Root Regions of Peanut Plant Under Greenhouse Condition

Sound mature peanut kernels were surface-disinfested by agitating in 20% household bleach and drying at 25 °C for 1 h in open petri dishes in a laminar flow hood. Surface-disinfested kernels were germinated in 3 l pots containing promix, moistened and autoclaved twice before use. At 7 days after emergence of plant from promix (at two leaves stage), root region of each plant was inoculated by known inoculum of bacteria and toxigenic *A. flavus* strains with an adjustable pipette having a disposable tip. In co-inoculation experiment, root region was inoculated simultaneously with toxigenic *A. flavus* and selected bacterial strain. In pre-inoculation experiment, bacterial strain was inoculated 1 day earlier inoculation of toxigenic *A. flavus* strain. Similarly, in post-inoculation experiment, bacterial strain was inoculated 1 day after inoculation of toxigenic *A. flavus* strain. In all the tests, peanut pods were harvested at maturity (14 weeks after inoculation) and dried at 60 °C for 48 h. Dried kernels were kept at room temperature in sealed plastic bags. All the experiments were performed twice, with six replicates for each treatment in each experiment.

Aflatoxin Extraction and Quantification

Aflatoxin was extracted by reversed-phase high-pressure liquid chromatography (Stubblefield and Shotwell 1977). The mobile phase consisted of HPLC grade acetonitrile: tetrahydrofuran: water (10: 6: 84, v/v/v), adjusted to pH 3.9 with acetic acid. The flow rate was 2 ml/min.

Results and Discussion

Seed Assay

The concentration of *A. flavus* inoculum had a significant effect on fungal growth in both the seed and root-radicle assays. Increasing the inoculum concentration of *A. flavus* from 25 spores/ml to log 6 spores/ml resulted in a 20 h difference in time to complete coverage of seeds by mycelia (Table 1).

An inoculum concentration of log 6 conidia per seed was selected for routine screening, since this gave complete coverage of seeds by mycelia within 4 days, which was deemed a good time frame for screening. It could therefore be inferred that the complete coverage of the seed by mycelia was dependent of the amount of inoculum. In these studies, the cultures developing from dilute inocula hardly sporulated even after a prolonged incubation; however, sporulation was much faster and more profuse in the cultures developing from greater inoculum levels, suggesting the involvement of mycelia differentiation in aflatoxin production.

Table 1 Effect of *A. flavus* inoculum density on fungal development of seeds

Inoculum density of *A. flavus* (spores/ml)	Hrs[a] to first appearance of fungus	Hrs[a] to complete coverage of seeds by the fungus
Log 6	60.62	97.25
Log 5	80.50	108.50
Log 4	82.50	109.50
Log 3	85.50	109.75
Log 2	89.16	110.16
50	92.23	115.66
25	93.20	117.20

[a]Mean value of eight replicates

Table 2 Effect of *A. flavus* inoculum density on fungal development of root radicles

Inoculum density of *A. flavus* (spores/ml)	Hrs[a] to first appearance of fungus	Hrs[a] to complete coverage of radicles by the fungus
Log 6	51.80	72.00
Log 5	54.57	75.00
Log 4	62.00	81.20
Log 3	76.00	99.60
Log 2	82.00	114.00
50	–	–
25	–	–

[a]Mean value of eight replicates

Root-Radicle Assay

More effects of inoculum concentration were noted in the root-radicle assay (Table 2) than the seed assay. Inoculum concentrations of 25 and 50 spores/ml were ineffective in the development of fungus on the root radicle, and log 6 conidia per radical was selected as the rate for routine screening. With this rate, mycelia were first apparent approximately 3 days after inoculation. Results of screening biocontrol agents in the root-radicle assay reveal that some bacteria completely prevent fungal growth, while others allow limited growth but prevent complete mycelia coverage of inoculated radicles. In addition, pronounced root growth enhancement was also noticed with some bacterial strains.

In these assay experiments with seven geocarposphere bacteria with known biocontrol activity against toxigenic *A. flavus* on peanut, four were found effective against growth and sporulation of *A. flavus*.

Antibiosis Test of Bacteria Against *A. flavus*

All seven bacterial strains mentioned above were screened for antibiosis towards *A. flavus* using two different media (PAF and TSA). Two bacterial strains showed a minimum inhibitory effect (0 to 1 mm), while one bacterial strain had maximum effect (5–10 mm) on PAF medium. However, on TSA medium, maximum zones of inhibition (10–12 mm) were observed with four bacterial strains (Table 3). Thus TSA medium appeared to be a suitable medium for testing antibiotic effect of different types of bacteria on *A. flavus*.

Inoculation of Bacteria/Toxigenic *A. flavus* in the Root Regions of Peanut Plant Under Greenhouse Condition

Co-inoculation experiments with toxigenic *A. flavus* and geocarposphere bacteria resulted in lower synthesis of Afl-B_1 in the peanut kernels as compared to control.

Table 3 Number of bacterial strains exhibiting antibiosis against *A. flavus*[a] on culture media

	Zone of inhibition (mm)				
	0–1	1–3	3–5	5–10	10+
PAF	2	2	2	1	0
TSA	0	0	1	2	4

[a]Based on evaluation of seven bacterial strains from geocarposphere of peanut plants

Table 4 Effect of geocarposphere bacteria on Afl-B_1 production in peanut kernels by toxigenic *A. flavus* strain

Organisms	Afl-B_1 content[a] of peanut kernels (μg/g)			
	Inoculated alone	Co-inoculated Au-32 and bacteria	Inoculated bacteria 1 day earlier Au-32	Inoculated bacteria 1 day after Au-32
Toxigenic *A. flavus* (Au-32) control	48.20	–	–	–
A.F./*B. megaterium*	–	36.60	3.47	30.50
A.F./*B. laterosporus*	–	230.90	9.12	35.10
A.F./*Cellulomonas cartae*	–	98.65	40.07	42.19
A.F./*Flavobacterium odoratum*	–	51.10	35.40	40.10
A.F./*Phyllobacterium rubiacearum*	–	75.65	37.16	40.80
A.F./*Pseudomonas aurofaciens*	–	57.15	15.50	37.20
A.F./*Xanthomonas maltophila*	–	50.45	19.17	38.15

[a]Mean value of eight replicates

Of seven bacterial strains tested, four strains showed a reduction in aflatoxins production in varying extents (Table 4). One day earlier inoculation of bacterial strains resulted in greater reduction in aflatoxin accumulation. *Bacillus megaterium* showed maximum inhibition of aflatoxin biosynthesis, as compared to remaining three potential bacterial strains.

In greenhouse experiments, significant level of Afl-B_1 (48.20 μg/g) was produced in peanut kernels when toxigenic strains of *A. flavus* were inoculated alone. A mixed type of results in Afl-B_1 production potential of *A. flavus* (i.e. highest, lowest and moderate level) were found in three different co-inoculation experiments with bacteria. When *A. flavus* was inoculated simultaneously with each of the seven bacterial strains, Afl-B_1 production increased in the case of all bacterial strains except *B. megaterium*. Interestingly, a fivefold increase in Afl-B_1 level (230.90 µg/g) was found when *A. flavus* was co-inoculated with *B. laterosporous*. Similarly, two- and 1.5-fold increases in Afl-B_1 level were noticed with *C. cartae* and *P. rubiacearum*, respectively. In pre-inoculation experiment, when bacterial strains were inoculated 1 day earlier of *A. flavus* inoculation, four bacterial strains showed a greater reduction in aflatoxin production in peanuts. The maximum reduction or negligible amount of Afl-B_1 (3.47 µg/g) in peanut kernels was found with *B. megaterium*. Similar reduction in Afl-B_1 level was recorded with *B. laterosporous* (9.12 µg/g), *P. aurofaciens* (15.5 µg/g) and *X. maltophilia* (19.17 µg/g). In post-inoculation experiment, i.e. when bacterial strains were inoculated 1 day after inoculation of *A. flavus*, Afl-B_1 level was not much reduced with all seven bacterial strains.

The present results of co-inoculation experiments indicate that almost all geocarposphere bacteria, except *B. megatorium*, stimulated the overall level of Afl-B_1 production by *A. flavus*. The 5-fold, 2-fold and 1.5-fold increases in Afl-B_1 production with *B. laterosporous*, *C. cartae* and *P. rubiacearum*, respectively, are might be due to the production of certain metabolites by these bacteria which altered the substrate, enhancing the growth of *A. flavus* or its ability to produce aflatoxin or both. It is possible that these bacterial strains could here enzymatically changed the substrates, making them more favourable for aflatoxin biosynthesis. Similarly, compounds from these organisms could have been released, leading to enhanced growth or aflatoxin production or both. It is also possible that these bacterial strains could attach to the hyphae of *A. flavus* and travelled with hyphae into the tissue, thereby altering not only the substrate but also the fungal wall membrane structure, changing its diffusibility or increasing the rate at which metabolites, such as the aflatoxins, can leak through the membrane.

The results of pre-inoculation experiments show that out of seven bacterial strains, two strains viz. *B. megaterium* and *B. laterosporous* reduced Afl-B_1 synthesis by 90–100%. The reduction in aflatoxin biosynthesis might be due to the nutritional competition, altering the levels of O_2 or CO_2 between the bacterial and toxigenic *A. flavus* strains, thereby exhibiting pronounced changes in growth, sporulation and toxin production. The degradation, detoxification and absorption of toxin by bacterial strains can cause a reduction in aflatoxin level. Bacterial cells can degrade the aflatoxin produced by *A. flavus*, possibly via production of cell wall lytic enzymes.

Conclusion

The results suggest the potential of bacteria in inhibition of pre-harvest aflatoxin biosynthesis in developing peanuts. Detailed investigations are desirable to bring the advantages of this greenhouse experiment to a field level and to evaluate the practicability of the approach. These biocontrol agents could be exploited for sustainable development in peanut.

Acknowledgements The authors are thankful to Prof. Barry J. Jacobsen, Head, Department of Plant Pathology, Alabama Agricultural Experiment Station, Auburn University, USA, for providing laboratory facilities. Financial support to HKC by Govt. of India, Dept. of Education, New Delhi in the form of Postdoctoral Fellowship is gratefully acknowledged.

References

Barnett HL, Hunter BB (1972) Illustrated genera of imperfect fungi. Burgress Publishing Company, Minneapolis, MN, USA

Bowen KL, Backman PA (1989) Effectiveness of fungicides for control of mycotoxigenic fungi and mycotoxins in peanut. Phytopathology 79:1188

Chourasia HK (1995) Kernel infection and aflatoxin production in peanut (*Arachis hypogoea* L.) by *Aspergillus flavus* in presence of geocarphosphere bacteria. J Food Sci Technol 32(6):459–464

Chourasia HK (2001) Response of some Indian maize samples for aflatoxin production by *Aspergillus flavus* strains. J Food Sci Technol 38(4):387–389

Chourasia HK, Suman SK, Pramila P (2008) Mycotoxigenic fungi and mycotoxins in fast foods of Bihar. J Mycol Plant Pathol 38(3):492–499

Dorner JW, Cole RJ, Blankenship PD (1990) The use of a biocompetitive agent to control preharvest aflatoxin in drought stress peanuts. In: Porter DM, Dickens JW (eds) Proceedings of the American peanut research and education society, vol 22. Inc, Yaokum, Texas, pp 34–39

Johnson JW, Curl EA (1972) Methods for research on ecology of soil-borne plant pathogens. Burgess Publishing Co., Minneapolis

Kloepper JW, Bowen KL (1991) Quantification of geocarposphere and rhizosphere effect of peanut (*Arachis hypogaea* L.). Plant Soil 136:103–109

Mandal NL, Singh AN, Chourasia HK, Roy AK (2010) Aflatoxin production potentiality of *Aspergillus flavus* strains associated with maize rhizosphere. J Mycopathol Res 48(1):91–93

Stubblefield RD, Shotwell OL (1977) Reverse phase analytical and preparative high pressure liquid chromatography of aflatoxins. J Assoc Off Anal Chem 60:784–790

Developing Efficient Methods for Unravelling Headspace Floral Volatilome in *Murraya paniculata* for Understanding Ecological Interactions

Ishita Paul, Priyal Goyal, Pratapbhanu Singh Bhadoria and Adinpunya Mitra

Abstract

Volatile compounds emitted by plants are important for understanding their ecological interactions. The volatile emission profile of a plant or plant part can be analysed by sampling the headspace air—an enclosed volume of air surrounding the plant or plant part. Polymeric or carbon-based adsorbents are often used to capture headspace volatiles, which are subsequently eluted using appropriate organic solvents. However, any adsorbent or solvent may have differential affinities for various volatile compounds. Five different adsorbents and two solvents were tested for sampling the floral headspace volatiles of *Murraya paniculata* (L.) Jack. These flowers emit scent volatiles of at least four biochemical classes and attract a diverse range of pollinators. Based upon the results, carbon-based absorbents were inferred to have higher affinities for high molecular weight, non-polar aliphatic volatiles. Commercially manufactured polymeric absorbent of Porapak Q series (ethylvinylbenzene-divinylbenzene) had higher affinities for terpenoid and slightly polar aliphatic compounds, whereas Tenax TA (2,6-diphenyl-p-phenylene oxide) granules captured aromatic ring-based volatiles more efficiently. Between the two organic solvents, hexane showed significantly higher and lower affinities for aliphatic and aromatic ring-based volatiles, respectively, than dichloromethane. There were, however, some compounds of intermediate polarities and molecular weights which proved

I. Paul · P. Goyal · A. Mitra (✉)
Natural Product Biotechnology Group, Agricultural and Food Engineering Department, Indian Institute of Technology Kharagpur, Kharagpur 721302, West Bengal, India
e-mail: adin@iitkgp.ac.in

I. Paul · P.S. Bhadoria
Soil Science & Plant Nutrition Laboratory, Agricultural and Food Engineering Department, Indian Institute of Technology Kharagpur, Kharagpur 721302, West Bengal, India

K. Mukhopadhyay et al. (eds.), *Applications of Biotechnology for Sustainable Development*, DOI 10.1007/978-981-10-5538-6_10

to be exceptional to these broadly defined categories. Moreover, often the qualitative diversity of the volatile profile captured was compromised for the quantitative efficiency of capturing volatiles by any given absorbent–solvent combination. It was concluded that the system of capturing of the volatile profile of a plant species should be standardized according to the biochemical nature of its volatilome.

Keywords

Headspace · Plant volatiles · Terpenoids · Benzenoids · Nitrogenous aromatic volatiles · Aliphatic volatiles

Introduction

Plant volatile organic compounds (PVOCs) are generally low molecular weight (<300 Da), lipophilic or amphiphilic compounds with low boiling points, which easily diffuse through living membranes and are thus easily emitted by plant tissues into the atmosphere (Pichersky et al. 2006). Over 1700 PVOCs are known, produced in various parts of plant species belonging to about 90 families of gymnosperms and angiosperms (Knudsen and Gershenzon 2006). These are mostly secondary metabolites formed by various metabolic pathways (Dudareva and Pichersky 2000). All these volatiles create a network of chemical communication which mediates countless interactions of the plants with their environments (Loreto et al. 2014). The PVOCs are also the source of the aromas and flavours of flowers, fruits, herbs and spices.

Traditional methods for extraction of PVOCs, such as hydro-distillation and solvent extraction, involve maceration or heating of the plant material, which lead to contamination with nonvolatile compounds, and typically have low yield. The exact profile of PVOCs emitted by small plant parts or even whole plants is captured efficiently in a "headspace"—an enclosed volume of air surrounding the plant. Air saturated by emitted PVOCs is sampled by packed beds of adsorbent material, which can retain very low quantities of specific PVOCs. Quantities of specific PVOCs, however, are affected by the composition of the adsorbent. Here we investigated the efficiency of a few such adsorbents and two solvents (for elution) in capturing the fragrance composition of flowers of *Murraya paniculata* (L.) Jack, and attempted to develop an effective combination for the same purpose.

Table 1 Adsorbent materials used for floral headspace trapping

Adsorbent material	Chemical composition	Additional information
Porapak™ Q	Ethylvinylbenzene-divinylbenzene	Two mesh sizes (50/80 and 80/100) used
Tenax® TA	2,6-diphenyl-p-phenylene oxide	Cannot be desorbed with dichloromethane
Charcoal	Carbon	Replaced with graphite in mixed adsorbents due to high volume of solvent required for desorption

Materials and Methods

Plant Material

Fresh crepuscular/nocturnal *M. paniculata* flowers, which reportedly emit benzenoids, C_2 phenylpropanoids, nitrogenous aromatics, monoterpenoids, sesquiterpenoids and aliphatic esters (Rout et al. 2007), were used for as source of PVOCs.

Adsorbent Materials

Commercially manufactured polymeric adsorbents, Porapak™ Q and Tenax® TA, and one carbon-based adsorbent were used separately for trapping of floral PVOCs (Table 1). On the basis of the results obtained, combinations of one polymeric adsorbent with a carbon-based adsorbent were prepared and used for a second set of trapping experiments.

Set-up for Floral Headspace Trapping

PVOCs were collected from floral material (250 mg) inserted in a 500 mL headspace following an experimental set-up modified from Bera et al. (2015). Saturation was allowed for 2 h, followed by collection by adsorbent (30 mg) packed beds for 20 min. The VOCs, filtered by the packed bed of adsorbents, were flushed out with an organic solvent (*n*-hexane or dichloromethane).

Analysis of VOCs by Gas Chromatography Mass Spectrometry (GC/MS)

All the samples were analysed on a Shimadzu QP2010SE GC/MS system. A ZB-5 column (30 m × 0.25 mm internal diameter, film thickness 0.25 μm) was used for

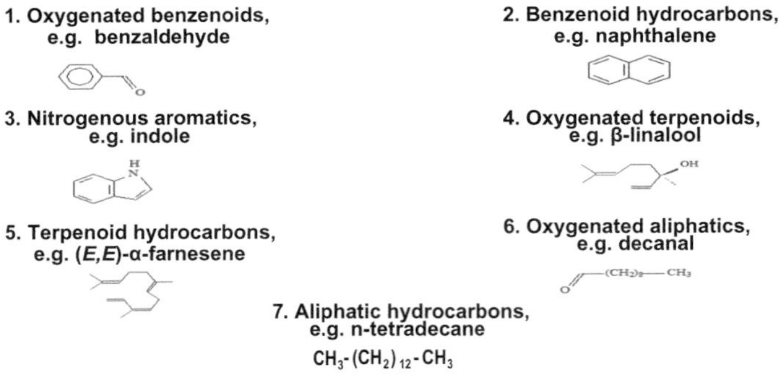

Fig. 1 Classification of PVOCs obtained by headspace trapping of *M. paniculata* flowers

the separation of volatile compounds, following the method of Maiti et al. (2014), with helium as a carrier gas. The results were processed by GC/MS solution (ver. 2.6 software).

Results and Discussion

Variety of PVOCs Emitted by *M. paniculata* Flowers

The diversity of PVOCs captured by the various adsorbent–solvent trapping systems from floral headspace of *M. paniculata* could be broadly classified into seven categories according to the nature of C–C bonds and functional groups (Fig. 1).

Trapping with Separate Adsorbent Materials

Relative quantities of different PVOCs as trapped by individual adsorbent materials varied widely (Fig. 2). It was evident that Porapak™ Q adsorbents had higher affinity for terpenoid and aliphatic hydrocarbons, and lower affinity towards oxygenated benzenoids and nitrogenous aromatics than Tenax® TA. However, relative PVOC quantities as captured by Porapak™ Q 50/80 showed very high variances, and the adsorbent was not used in subsequent experiments. Charcoal was the originally tested carbon-based adsorbent, but was found to require large volumes of desorbing solvent due to its high retentive capacity, resulting in extremely dilute samples. Graphite, which was similar to charcoal in showing greater affinity for aliphatic hydrocarbons than any other PVOC categories, was used as carbon-based adsorbent material in subsequent experiments.

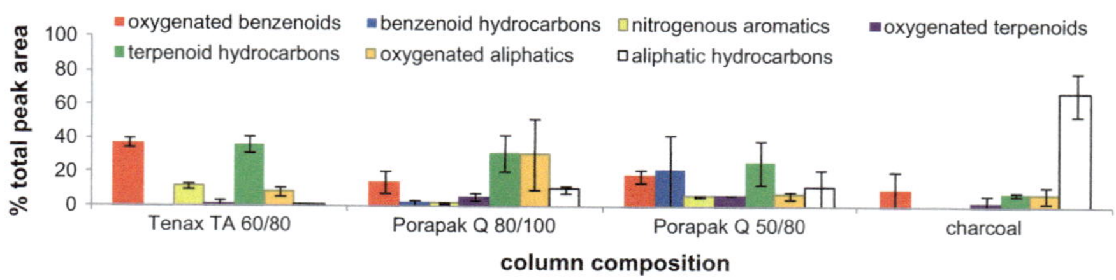

Fig. 2 PVOC profiles of *M. paniculata* floral fragrance as captured by different individual adsorbent materials and eluted with *n*-hexane

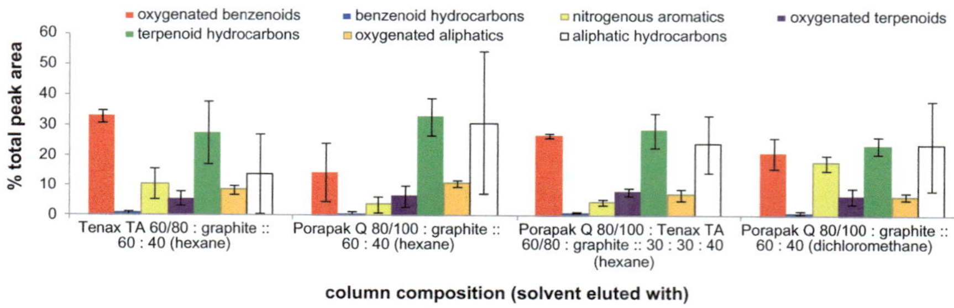

Fig. 3 PVOC profiles of *M. paniculata* floral fragrance as captured by mixed adsorbent materials and eluted with two different solvents

Trapping with Combined Adsorbent Materials

Combination of polymeric adsorbent materials with carbon-based adsorbent obtained more similar PVOC profiles than achieved by unmixed adsorbents (Fig. 3). After desorption with *n*-hexane, PVOC profiles as trapped by Porapak™ Q 80/100 + graphite mix (60:40) consisted of 4–25% oxygenated benzenoids, 1–7% nitrogenous aromatics and 27–39% terpenoid hydrocarbons; whereas the same PVOC groups as captured by Tenax® TA 60/80 + graphite (60:40) mix constituted 30–36, 5–16 and 17–38%, respectively. Interestingly, triple adsorbent mix of Porapak™ Q 80/100 + Tenax® TA 60/80 + graphite (30:30:40) captured these three PVOC groups at intermediate values. However, proportions of oxygenated terpenoids and oxygenated aliphatics were, respectively, higher and lower in fragrance profiles trapped by the triple mixture as compared to both the dual mixtures. Proportions of aliphatic hydrocarbons varied extremely, but the mean and variance were less extreme as captured by the triple mixture; while those of benzenoid hydrocarbons were negligible in most samples.

Desorption Efficiency of *n*-Hexane and Dichloromethane

Desorption of the dual adsorbent mixture of Porapak™ Q 80/100 + graphite was done with either *n*-hexane or dichloromethane. It was inferred from the resulting PVOC profiles that dichloromethane has greater affinity for oxygenated benzenoids and nitrogenous aromatics, and lesser affinity for terpenoid hydrocarbons, oxygenated hydrocarbons and aliphatic hydrocarbons, than *n*-hexane (Fig. 3). Both solvents showed similar preferences for oxygenated terpenoids.

Qualitative Diversity of PVOCs

Qualitative nature of the fragrance, as measured by total number of PVOCs trapped, was captured most efficiently by the triple adsorbent bed consisting of Porapak™ Q 80/100 + Tenax® TA 60/80 + graphite which trapped 38–50 PVOCs. The least PVOC diversity was captured by individual carbon-based adsorbents, which could trap only 15–21 compounds. Between the desorbing solvents, dichloromethane could elute more (33–40) volatiles than *n*-hexane (30–36).

Conclusion

Exact quantification of PVOCs in *M. paniculata* floral fragrance was not achieved because of the variations in volatile profiles obtained by different combinations of adsorbents and desorbing solvents. However, common features of PVOC profiles captured by mixed adsorbent beds included high proportions of oxygenated benzenoids, terpenoid hydrocarbons and aliphatic hydrocarbons. It may safely be concluded that these features are true for the composition of *M. paniculata* floral fragrance. These observations also indicate that PVOC composition of any fragrant plant species should be ascertained by more than one combination of adsorbents and solvents if headspace method is to be used. Using only one combination for method optimization might lead to inadvertent bias towards specific PVOCs.

Acknowledgements Ishita Paul is grateful to the Council of Scientific and Industrial Research (CSIR) for awarding a Shyama Prasad Mukherjee (SPM) Fellowship for pursuing a doctoral degree at IIT Kharagpur. This work was supported by a grant (EMR/2015/000247/PS) from the Science and Engineering Research Board (SERB), India.

References

Bera P, Kotamreddy JNR, Samanta T, Maiti S, Mitra A (2015) Inter-specific variation in headspace scent volatiles composition of four commercially cultivated jasmine flowers. Nat Prod Res 29(14):1328–1335

Dudareva N, Pichersky E (2000) Biochemical and molecular genetic aspects of floral scents. J Plant Physiol 122(3):627–633

Knudsen JT, Gershenzon J (2006) The chemical diversity of floral scent. In: Dudareva N, Pichersky E (eds) Biology of floral scent. CRC Press, Boca Raton, pp 27–52

Loreto F, Dicke M, Schnitzler JP, Turlings TCJ (2014) Plant volatiles and the environment. Plant Cell Environ 37(8):1905–1908

Maiti S, Moon UR, Bera P, Samanta T, Mitra A (2014) The in vitro antioxidant capacities of *Polianthes tuberosa* L. flower extracts. Acta Physiol Plant 36(10):2597–2605

Pichersky E, Noel JP, Dudareva N (2006) Biosynthesis of plant volatiles: nature's diversity and ingenuity. Science 311(5762):808–811

Rout PK, Rao YR, Sree A, Naik SN (2007) Composition of essential oil, concrete, absolute, wax and headspace volatiles of *Murraya paniculata* (Linn.) Jack flowers. Flavour Fragr J 22:352–357

Studies on Nutraceutical Properties of *Annaona squamosa*

S. Bala, V.K. Nigam, A.K. Tiwari and A.S. Vidyarthi

Abstract

Annaona squamosa Linn. is a highly important plant with edible and medicinal values. Custard apple (*Annaona squamosa Linn.*) is considered as a well-balanced food which contains about 43% of edible part consisting of 1.60% protein, 21% sugars, 23.90% carbohydrates, 74.60% moisture, calcium, phosphorus, iron, etc. and with a total calorific value of 105 calories per 100 g of edible pulp. Besides these nutritional components it also contains a range of pharmaceutically important constituents like alkaloids, acetogenines, cyclopeptides, kaurane diterpinoids, etc., that can act as nutraceuticals. It has been reported that there is a huge postharvest loss of custard apple in various parts of country and therefore several products from custard apple have been under the process of development. In this study, an attempt has been taken to develop jam from the custard apple pulp.

Keywords

Annona squamosa · Ascorbic acid · Minerals · Jam · Nutraceuticals

S. Bala · V.K. Nigam (✉) · A.S. Vidyarthi
Department of Bio-Engineering, Birla Institute of Technology, Mesra, Ranchi 835215, India
e-mail: vknigam@bitmesra.ac.in

A.K. Tiwari
Department of Chemical and Polymer Engineering, Birla Institute of Technology, Mesra, Ranchi 835215, India

K. Mukhopadhyay et al. (eds.), *Applications of Biotechnology for Sustainable Development*, DOI 10.1007/978-981-10-5538-6_11

Introduction

The genus Annona, of the family Annonaceae, comprises 60 or more species, typically of tropical American origin (Morton 1966). The four well-recognized bearers of edible produce are true custard apple or bullock's heart (*A. reticulata Linn.*), the sugar apple or sweetsop (*Annona squamosa Linn.*), the cherimoya (*A. cherimola Linn.*), and the sour sop (*A. muricata Linn.*) (Morton 1946. Among these *A. squamosa L.* fruit is called as Buddha's head after its shape in India, and are commonly known as custard apple, sugar apple, sweet sop or Atemoya in other part of the world. In India Custard apple (*A. squamosa* L.) is also popularly called as sitaphal, seureuba, sugar apple, gishta, sweet apple, zimtapfel, sharifa, sitappalm, matomoko, aati, atis, and anoda, is a tropical branched shrub cultivated throughout India for its fruits and different parts viz. seed, leaf, root, bark for the treatment of various diseases (Padhi 2011).

In India, custard apple is grown in Maharashtra, Gujarat, Andhra Pradesh, Uttar Pradesh, Madhya Pradesh, Bihar, Jharkhand, Assam, Rajasthan, Orissa and Tamil Nadu (APEDA). The custard apple was produced as 20,497 MT from 4,990 ha of area in Maharashtra (Sontakke 2003). Its cultivation in India was estimated to be 30,000 ha with an annual production of 228 thousand MT in the year 2014–15 while for the year 2015–2016 cultivated in 35,000 ha area with production of 271 thousand MT (NHB report 2015–16).

Custard apple is a nutritionally balanced food which constitutes protein, fiber, minerals, vitamins, energy, and very little fat (Bose 2002). It gives about 40% pulp with TSS 26.4° brix, pH 5.5 and tannins 0.5%, and may play an important role as nutraceuticals (Bhardwaj et al. 2014). It is an excellent store of vitamin C, vitamin B2, vitamin B6, vitamin A, a good source of dietary fiber, magnesium, and potassium which can improve fortitude and maintain good health (Davidson et al. 1975; Vaidya et al. 2016).

Storage of custard apple fruits has limitations due to its very short shelf life and chilling injury at low temperature (Chadha 2006; Prasanna 2000). In these circumstances, there is need to develop standard techniques to reduce those postharvest losses (Chikhalikar et al. 2000). Food processing is very important to minimize postharvest loss and to improve linkages between industry and agriculture (Gamage et al. 1997; Chaoba Singh 2000). Jam processing is a type of fruit preservation method that adds value to overripe fruits and prevents loss of such valuable nutrients and ensures nutritional security. The main objective for carrying out this research was to utilize custard apple in the preparation of jam and analysis of various physico chemical characteristics.

Materials and Methods

Collection and Sorting of Custard Apple

Custard apple (*A. squamosa L.*) fruits were procured from the local market of Ranchi district (Jharkhand). Fresh and fully ripe fruits were selected and sorted to remove diseased, bruised, rotten, bird-eaten fruit. Fruits were washed with running tap water, treated with a solution of sodium hypochlorite (200 ppm) for 15 min followed by thorough washing with water.

Extraction of Pulp from Custard Apple

Selected custard apple fruits were cut into two halves and the pulp was scooped with stainless steel spoon. Subsequently, the pulps were passed through pulping machine to separate seeds from pulp. The resulting pulp was mixed with 1000 ppm of KMS (Potassium meta bi sulphite) and stored for further processing.

Physico—Chemical Analysis of Pulp and Jam

Physical parameters like weight of fruits, weight of pulp, weight of skin and weight of seed were measured by recording weight of each sample on a weighing balance (German Make) with 0.001 g least count and percentage value of each of the parameter were reported.

The chemical constituents of pulp and jam were analyzed to assure the quality of Jam. Total Soluble Solids of jam was determined by using ERMA hand refractometer and expressed in °B, the titrable acidity and ascorbic acid content of pulp and jam were estimated by the methods described by Ranganna (1986). Sugars of pulp and jam were determined by Lane Eynon method (Haq and Rab 2012). The pH of pulp and jam were determined by Zenieth digital pH meter. The moisture content and Total nitrogen were determined following standard methods of analysis (AOAC 2006) and ash determination by (AOAC 2003).

Preparation of Custard Apple Jam

Custard apple pulp was used for preparation of jam in an open steel kettle as described by Awan and Rehman (1999). About 80 g of custard apple pulp was mixed with 20 ml of water, 50 g of sucrose were added and the mixture was heated for about 3–5 min. Commercial food-grade pectin (0.75 g) was mixed in hot water until dissolved separately to prevent lump formation and added to the kettle for cooking to about 65 °B with manual stirring. TSS (°B) of the reaction contents was determined at 25 °C. About 0.5 g of citric acid was added at the end of cooking to adjust pH around 3.2–3.4. The jam thus prepared were poured into 200 g sterile glass jars, packed and cooled to room temperature. Knife test was done to check the jam setting quality.

Results and Discussion

Physical Characteristics of Custard Apple

Annonas belongs to a class of underutilized fruit categories that contain a huge source of the components for medicines and other industrial products (Pandey and Barve 2011 and Bing et al. 2015). The various physical parameters of custard apple are shown in Table 1.

The color of custard apple fruit is medium green, becoming light green or greenish yellow at maturity, with a white, powdery bloom. The pulp is white, edible with sweet pleasant aroma. The starting material of single custard apple fruit weight was 186 g. The Number of seeds, total weight of pulp, fruit to pulp ratio and pulp to seed ratio recorded were 43, 79.98 g and 2.32, respectively. The number of seeds and weight of seeds per fruit was 43 and 12.3 g respectively.

Table 1 Physical parameters of custard apple

S. No.	Parameter/property	Mean[a] ± SE
1.	Weight of fruit (g)	186 ± 14.97
2.	Peel (g)	85.32 ± 7.23
3.	Peel (%)	45.69 ± 0.92
4.	Fruit pulp (g)	79.98 ± 6.3
5.	Fruit pulp (%)	43 ± 3.79
6.	Number of seeds per fruit	43 ± 3.86
7.	Seeds per fruit (g)	12.30 ± 1.26
8.	Seeds per fruit (%)	6.54 ± 0.31
9.	Fruit to pulp ratio	2.32 ± 0.04
10.	Length of fruit (cm)	5.4 ± 0.3
11.	Breadth of fruit (cm)	4.3 ± 0.2
12.	Waste index (%)	58.42

[a]Mean of 10 sample ± standard deviation of samples

Table 2 Chemical composition of 100 g of edible custard apple pulp

S. No.	Chemical parameters	Content
1.	TSS (°B)	28
2.	Reducing Sugar (g)	20.15
3.	Total sugar (g)	20.95
4.	Carbohydrate (g)	23.05
5.	Vitamin C (mg)	32.5
6.	Ash (g)	0.6
7.	Titrable acidity (g)	0.29
8.	pH	5.48
9.	Moisture %	74.6
10.	Protein (g)	1.63
11.	Calorific value (Kcal)	105

Table 3 Chemical composition of custard apple jam for 100 g of pulp

S. No.	Chemical parameters	Content
1.	TSS (°B)	68
2.	Moisture %	26.23
3.	Reducing Sugar (g)	21.6
4.	Total sugar (g)	51.36
5.	Carbohydrate (g)	74.3
6.	Titrable acidity (g)	0.43
7.	Protein (g)	0.37
8.	Calorific value (Kcal)	295.03
9.	Ash (g)	0.13
10.	pH	3.48

Chemical Characteristics of Custard Apple Pulp and Jam

The chemical characteristics of custard apple pulp and jam are given in Tables 2 and 3.

Preparation of Jam

The jam was prepared as described in Section "Materials and Methods". It was observed during preparation of jam that pectin is required in the ratio of 0.75–1% for proper setting of jam which indicated less amount of available pectin in custard apple fruits.

The table indicates low moisture content of jam and had soluble solids mean values of 68.00 °Brix. Sugar in jams contributes to high content of soluble solids, an effect that is essential for the physical, chemical and microbiological stability and possible gelation. The moisture content of jam reduced to 26.23% as compared to pulp which was 74.6%. The studies showed that carbohydrate content of pulp and jam were 23.05 and 74.3 g respectively. The processed jam showed presence of reducing sugar as 21.6 g, titrable acidity 0.43 g, protein 0.37 g and pH of 3.48.

Nutraceutical Role of Custard Apple Pulp and Jam

Custard apple contains high calorific value of pulp and jam as 105 and 295 kcal per 100 g of edible portion respectively. Based on the calorific values of processed products, it can be included in diet for weight gain and athletes. Despite of high reducing sugar content 21 g, the glycemic index of custard apple pulp is low (54) and due to presence of high antioxidant content it is suitable for diabetic patient (Andrade et al. 2001).

A study showed that consumption of custard apple and processed products exhibited cardio protective effects like therapeutic doses of captopril used by heart patients. (Kaleem et al. 2006). Further, it has been observed that feeding custard apple pulp increased haemoglobin levels by up to 21% in animals (Gupta et al. 2005).

Conclusions

Processing of underutilized custard apple fruits will help in conserving its nutrient by preparation of various custard apple based products to increase nutritional value and accelerate the value addition. The study concluded the requirement of pectin for gelation of jam. The high content of sugar inhibits the growth of microbial contaminants and hence enhanced the storage and stability of the products. The processed product, i.e. jam can be industrially prepared with minimum cost as it preserves all the nutraceutically important constituents of fruits. Custard apple based products could enhance the conservation of other tropical fruits in the various value-added products.

References

Andrade E, Zoghbi M, Fabricius H, Marx F (2001) J Food Comp Anal 14(2):227–232

AOAC (2006) Official methods of analysis, 18th edn. AOAC, Gaithersburg

APEDA: http://apeda.gov.in/apedawebsite/trade_promotion/study_and_report.hm

Awan JA, Rehman SU (1999) Food preservation manual. Uni. Tech. Publishers, Faisalabad

Bhardwaj A, Satpathy G, Gupta RK (2014) Preliminary screening of nutraceutical potential of *Annona squamosa*, an underutilized exotic fruit of India and its use as a valuable source in functional foods. J Pharmacogn Phytochem 3(2):172–180

Bing H, Tong-Dan W, Shao-Ming S, Yun Y, Chan M, Zhu-Jun Y, Li-Shun W (2015). Annonaceous acetogenin mimic AA005 induces cancer cell death via apoptosis inducing factor through a caspase-3-independent mechanism. Hans et al BMC Cancer 15:139

Bose TK (2002) Fruits – Tropical and Subtropical, vol II. Kalyani Publishers, New Delhi, pp 293–328

Chadha KL (2006) Handbook of Horticulture ICAR Publication, New Delhi, pp 109–114

Chaoba Singh TH (2000) Seminar on Food processing and Nutritional objectives. Chall New Trade Order Assoc News 54:2–6

Chikhalikar NV, Sahoo AK, Singhal RS, Kulkarni PR (2000) Studies on frozen pourable custard apple (*Annona squamosa* L.) pulp using cryoprotectants. J Sci Food Agric 80:1339–1342

Davidson SR, Passmore JF, Brock AS (1975) Trustwell, human nutrition and dietics, 6th edn. Churchill Livingstone, pp 390–629

Gamage TV, Yuem CMC, Wills RBH (1997) Minimal processing of custard apple (*Annona atemoya*) pulp. J Food Process Preserv 21:289–301

Gupta RK, Kesari AN, Watal G, Murthy PS, Chandra R, Tandon V (2005) Ann Nutr Metabol 49:407–413

Haq IU, Rab A (2012) Characterization of physico-chemical attributes of litchi fruit and its relation with fruit skin cracking. J Anim Plant Sci 22(1):142–147

Kaleem M, Asif M, Ahmed QU, Bano B (2006) Singap Med J 47(8):670–675

Morton JF (1966) Florida state horticultural society. University of Miami, Coral Gables, pp 355–366

Morton K, Morton JF (1946) Fifty tropical fruits of Nassau. Text House, Coral Gables, pp 95–106

NHB (National Horticulture Board) Area and production of horticulture crops—2015–16 (3rd Advance Estimates)

Padhi LP, Panda SK, Satapathy SN, Dutta SK (2011) In vitro evaluation of antibacterial potential of *Annona squamosa* Linn. and *Annona reticulata* L. from Similipal Biosphere Reserve, Orissa, India. J Agric Technol 7(1):133–142

Pandey N, Barve D (2011) Phytochemical and pharmacological review on *Annona squamosa* Linn. Int J Res Pharm Biomed Sci 2(4):1404–1412

Prasanna KNV, Rao DVS, Krishnamurthy S (2000) Effect of storage temperature on ripening and quality of custard apple (*Annona squamosa* L.) fruits. J Hortic Sci Biotechnol 75(5):546–550

Ranganna S (1986) Handbook of analysis and quality control for of fruit and vegetable products, vol 12–21, 2nd edn. Tata Mc Graw-Hill Publishing Company Limited, New Delhi, pp 875–879

Sontakke MB (2003) Importance of rainfall fruit crops. Winter school on advances in production and management of rain fed fruit crops, pp 1–3

Vaidya A, Solanke ND, Gaware K (2016) Chemical composition, physicochemical and functional properties of custard apple (*Annona squamosa*) seed flours and protein isolate. Int J Sci Eng Technol 5(4):205–209

Automated Detection of Chronic Alcoholism Using Hilbert Huang Transformation

Surendra Kumar and Rakesh Kumar Sinha

Abstract

In this study, the magnitude and spatial distribution of frequency spectrum in the resting electroencephalogram (EEG) were examined to address the classification of alcoholism in the central nervous system. The EEG signals for chronic alcoholic conditions taken from motor cortex region were divided into five sub-frequency band and Hilbert Huang Transform is applied for feature extraction of the EEG signals. Since the extracted feature has large data dimension, it is reduced using Linear Discriminate Analysis. Support Vector Machine has been used for classification. Highest Classification accuracy 76.67% has been seen in the central (CZ) and left central (C3) and left parietal (P3) hemisphere when the classifier was tested with 150 EEG epochs of two seconds. In present results CZ, C3, P3 focal area of brain shown the better classification accuracy, which can be concludes as the persons with chronic alcoholism are having hyper active zone in comparison with control. Considering the classification accuracy obtained by SVM clustering with mean band power features in most of the EEG channels of motor cortex, it can be suggested that the noninvasive EEG signals can be considered as a tool for development of an automated online diagnostic system for the chronic alcoholic condition.

Keywords

Alcohol · Cerebral motor cortex · Electroencephalogram · HHT · LDA · SVM

S. Kumar (✉) · R.K. Sinha
Department of Bio-Engineering, Birla Institute of Technology, Mesra, Ranchi 835215, Jharkhand, India
e-mail: sikuranchi@gmail.com

K. Mukhopadhyay et al. (eds.), *Applications of Biotechnology for Sustainable Development*, DOI 10.1007/978-981-10-5538-6_12

Introduction

Alcohol has developed as one of the most costly and harmful addictions across the globe. Alcoholic beverages contain ethanol, a psychoactive drug with relaxants and euphoric effects, consumed by people throughout the world (Acharya et al. 2012). Alcohol abuse leads to various medical complications ranging from illness to disabilities which have led to a significant rise in the mortality rate over the period of time. Clinicians often face difficulty in identifying for problem of alcohol abuse in their patients as usually denial to the situation is hallmark in alcoholism (Kumar et al. 2015a, b). Chronic alcoholism has severe effects on cognitive and behavioral functions. Alcohol impairs the neural activity, with immediate effect on multiple cognitive motor processing domain. Magnetic Resonance Imaging (MRI) and Computed Tomography (CT) scans shown that the gray matter and white matter in the brain have reduced significantly (Crews et al. 2006; Calhoun et al. 2004) but they are costly diagnosis tools. It is therefore necessary to develop an automated tool for detecting chronic alcoholic subjects with noninvasive tools at low cost. EEG or the electroencephalogram are one of the solutions, where the signals that are picked up from the cerebral cortex of the brain. EEG signals represent the electrical activity of the brain generated from billions of neurons present. The EEG is one of the most complex physiological signals available which is used for evaluation of brain disorders. Usually it is used to show the type and location of the activity. The EEG offers unmediated deduction of the cortical behavior of the brain accompanied with millisecond temporal activity (Yeddula 2012).

In this study, our main aim was to find a permanent variation in a sample of EEG database and establish the variation as a consequence of alcohol abuse. In the present work spectral changes in EEG, if any, were being observed on the cerebral motor cortex and brain signals from 9 different electrode positions were acquired. The electrode positions being F3, Fz, F4, C3, Cz, C4, P3, Pz, and P4 electrodes. The alcoholic database was compared to a standard non-alcoholic or control group of EEG database and results were obtained. Additionally, Hilbert Huang Transform (HHT), an adaptive data analysis tool developed specially for analyzing nonlinear data was used along with an integrated framework of Linear Discriminate Analysis (LDA) and Support Vector Machine (SVM). Approximate entropy of both alcoholic and control subjects was also seen a distinguishing factor in this study.

Materials and Methodology

Subjects and Data Recording: The study was carried on a group of 40 male members with average age being 35 years. The group was equally divided into alcoholic and control subjects. The procedure of data recording was purely non-invasive and well explained to the subjects. Present research methodology was approved by the Institutional Ethical Committee of Rajendra Institute of Medical Sciences, Ranchi, India which follows the 'ethical guidelines for biomedical research on human subjects' provided by Indian Council of Medical Research, New Delhi. Further, written and signed consent were also taken from each of the individual subjects participated in the data recording. It is reported that the alcoholic subjects had been consuming Mahuwa, which is obtained and fermented from the flower of Madura longifolia. Alcoholic subjects had been consuming alcohol on a regular basis from past 10 years. The control subjects had never consumed alcohol or tobacco ever. EEG data was recorded using RMS System (Recorder & Medicare Systems Pvt. Ltd, India) of 19 Channel with Ag/AgCl electrodes placed using the International standard 10–20 electrode system.

Feature Extraction by Hilbert Huang Transform: Hilbert Huang Transform is the name designated by NASA for the combined framework of Empirical Mode Decomposition (EMD) and Hilbert Spectral Analysis (HSA) (Huang et al. 2008). HTT was exclusively developed for analysis for nonlinear and nonstationary records. Fourier analysis performs well when the system is linear and is not preferred when the data is nonstationary. Fourier analysis succeeded in providing frequency information of the analytic signal but fails to give any information about the occurrence of time (Rosen et al. 2014). EMD decomposes the nonlinear and nonstationary signals into its constituent components, known as the Intrinsic Mode Functions (IMF), which are usually small. The first IMF contains the highest frequency content and the final IMF component has the lowest. An IMF is defined as any function having same number of zero crossings and having symmetric envelopes defined by local maxima and local minima. Present decomposition method operates on time domain which is an adaptive in nature and therefore highly efficient. Since the decomposition is carried on the local characteristic time scale of the data, therefore it is suitable for nonlinear and nonstationary processes. HHT outperforms the traditional methods in the feature extraction procedure as it allows analyzing the heterogeneous signals in a similar framework based on the estimation of instantaneous amplitudes and frequencies (Zong and Chetouani 2009). It was found that HHT was more efficient than the other traditional methods which used for data analysis of nonstationary signals (Huang 1998).

Further, the dimension of feature excreted from HTT are reduced with the linear discriminate analysis, which is a method of finding linear combination among variables which best separates two or more classes. LDA searches for protrusions in which the data point of different classes is far away, while the data point of the same class is close together. (Kumar et al. 2015a, b; Ye et al. 2014). The reduced feature vector was classified with is a supervised learning model of SVM, which finds its basis from the statistical learning theory. Classification using SVM is achieved by mapping the original feature space into the high dimension feature space using a nonlinear function, represented as $\varphi(x)$. The regression function is given by $f(x) = \omega.\varphi(x) + b$; where ω the weight vector and b is being the bias. (Lin et al. 2008; Mantri et al. 2012; Shankar et al. 2014).

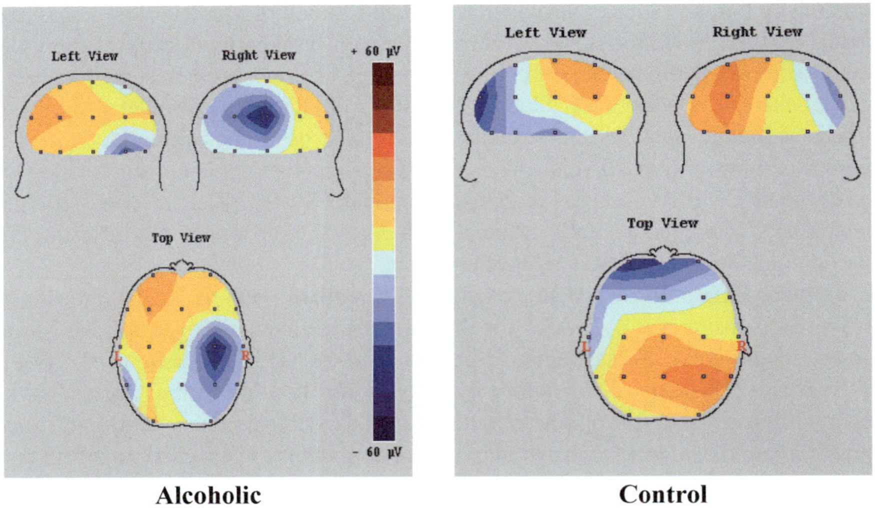

Fig. 1 Figures represent the comparison of distribution of EEG amplitude in left and right hemisphere of brain, between alcoholic and control subjects

Results and Discussions

In this study, we had proposed an innovative and a unique approach for detecting the condition for chronic alcoholism. The feature vector can be used as a distinguishing feature for gathering EEG into alcoholic and control groups. The comparison of distribution of EEG amplitude in left and right hemisphere of brain, between alcoholic and control subjects is presented in Fig. 1.

For the present study, band power of 400, 360, 320 and 300 EEG epochs of two seconds the data set has been divided equally into alcoholic and control groups. The IMF obtained from the EMD algorithm was used to obtain the feature vector. The IMF itself can be used as discriminating feature for classifying alcoholics and control subjects. The different number of samples of EEG data was classified using the combined framework of LDA and SVM.

The classification accuracies in percentage for the alcoholic and control subjects are presented in Table 1. Highest classification accuracy 76.67% has been seen in the central (CZ) and left central (C3) and left parietal (P3) hemisphere when the classifier was tested with 150 EEG epochs of two seconds. In present results, CZ, C3, P3 focal area of brain shown the better classification accuracy, which can be concludes as the persons with chronic alcoholism are having hyper active zone in comparison with control. It is also found that P3 is following the same pattern as of C3. We can thus conclude that the alcoholics have hyperactive central and parietal zones from these results. Therefore, it can be concluded that the left parts of motor cortex are more sensitive to chronic alcoholism and fail to synchronizing activity in

Table 1 Classification accuracy in percentage (%) for the alcoholic and control subjects EEG epochs of two seconds tested on different number of samples

Electrode position	Number of alcoholic and control EEG epochs selected			
	150	160	180	200
F3	60	60	58.33	60
Fz	70	65.25	61.11	62.5
F4	73.3	71.87	72	72.5
C3	76.6	75	75	75
Cz	76.6	71.875	72.22	70
C4	70	68.75	66.66	70
P3	73.33	71.87	72.22	65
Pz	76.67	71.78	72.22	72.5
P4	70	68.75	66.66	70

chronic alcoholism. Similar results were reported in the study of detection of chronic alcoholism using Fourier analysis (Kumar et al. 2015a, b). It has also been observed from the results that the odd numbered electrodes (F3, C3, P3) are having greater accuracy than the even numbered electrodes (F4, C4, P4). It is thus concluded that the left hemisphere of the brain is more vulnerable to the effects of chronic alcoholism than the control (nonalcoholic) subjects. Also, the *z*-line region had greater accuracy than the left hemisphere region. It was observed that on increasing the number of samples for classification the classification accuracy was reducing. It can be thus correlated that HTT greater accuracy for data analysis of small number of samples.

In present study, all the subjects were consuming alcohol for over 10 years and these are not the instantaneous result of alcohol. Alcoholism is seen as a complex physiological index and a wide range of biological variables are involved in the study of its symptoms such as physical dependence, craving, and impaired control. A few biochemical parameters along with the various neurological parameters are being studied to evaluate them as biological markers for detecting chronic alcoholism. In the chronic alcoholism, neurotransmitter systems like gamma aminobutyric acid (GABA) and glutamate are highly sensitive resulting in reducing overall brain excitability (Porjez and Begleiter 2003). In many alcoholic research studies, researchers also have reported about the shrinkage of brain tissues or neuronal loss in the superior frontal and motor cortices of the brain (Rangaswamy et al. 2002, 2004). With considering together, these findings suggest that alcoholics may account for their lack of CNS inhibition (i.e., hyper excitability) is caused due to deficit in GABA receptors in the brains of. (Rangaswamy et al. 2003).

The study could demonstrate significant alterations in the frequency spectrum of EEG signals for alcoholic subjects. But it is quite understood that these results based on the frequency spectrum will not be helpful foe the clinicians in their daily for detecting the problem of chronic alcoholism. Hence, future work may be planned to incorporate the proposed algorithm on an integrated digital chip which is quite easy task. The results obtained from the present study encourage the idea of development of a noninvasive automated identification system for detecting the problem of chronic alcoholism.

Conclusion

The main aim of this study is to develop a synchronized system for detecting alcoholism and observe the effect of alcohol on the brain waves. Band power is chosen as a major feature for studying the effect of alcohol in EEG. Hilbert Huang Transform was used to extract the features from alcoholic and control data. Linear Discriminate Analysis (LDA) and Support Vector Machine (SVM) were used to classify the EEG patterns into alcoholic and control groups. Maximum classification accuracy of 76.6% was obtained using the proposed technique. Further, the statistical analysis of band power of different EEG bands showed elevated levels for alcoholic subjects who can be seen as a marker for alcoholism or developing chronic alcoholism.

Acknowledgements Authors are grateful to Dr. Suhash Tetarway (Associate Professor, Department of Physiology, Rajendra Institute of Medical Sciences, Ranchi, India) for his support in EEG data recording and clinical study of alcoholism. Authors also acknowledge Mr. Pillutla Soma Sunder, M.Tech. Student, Department of Bio-Engineering, Birla Institute of Technology, Mesra, Ranchi (India) for the help received in the analysis of EEG data.

References

Acharya UR, Sree SV, Chattopadhyay S, Suri JS (2012) Automated diagnosis of normal and alcoholic EEG signals. Int J Neural Syst 22(3)

Calhoun VD, Altschul D, McGinty V, Shih RA, Scott D, Pearlson GD (2004) Alcohol intoxication effects on visual perception: an f-MRI study. Neuroimage 21:15–26

Crews FT, Mdzinarishvili A, Kim D, He J, Nixon K (2006) Neurogenesis in adolescent brain is potently inhibited by ethanol. Neuroscience 137:437–445

Huang N (1998) The empirical mode decomposition and the Hilbert spectrum for nonlinear and non-stationary time series analysis. Proc R Soc London 454:903–995

Huang NE, Wu Z, Long SR (2008) Hilbert huang transform. Scholarpedia 3(7):25–44

Kumar S, Ghosh S, Tetarway S, Sinha RK (2015a) Support vector machine and fuzzy C-mean clustering based comparative evaluation of changes in motor cortex electroencephalogram under chronic alcoholism. Med Biol Eng Comput

Kumar S, Ghosh S, Tetarway S, Sawai S, Sunder PS, Sinha RK (2015b) Non-linear autoregressive model for detecting chronic alcoholism. Computational intelligence in data mining, vol 2, Springer, Berlin, pp 393–402

Lin Y, Wang C, Wu T, Jeng S, Chen J (2008) Support vector machine for EEG signal classification during listening emotional music. In: 10th IEEE conference on multimedia signal processing, 8–10 Oct 2008, pp 127–130

Mantri S, Patil V, Mitkar R (2012) EEG based emotional distress analysis. Int J Eng Res Dev 4 (6):24–28

Porjez B, Begleiter H (2003) Alcoholism and human electrophysiology. Alcohol Res Health 27 (2):151–160

Rangaswamy M, Porjez B, Chorlian DB (2002) Beta power in the EEG of alcoholics. Biol Psychiat 51:831–842

Rangaswamy M, Porjez B, Chorlian DB (2003) Theta power in the EEG of alcoholics. Alcohol Clin Exp Res 27(4):607–615

Rangaswamy M, Porjez B, Chorlian DB (2004) Resting EEG in offspring of male alcoholics: beta frequencies. Int J Psychophysiol 51:239–251

Rosen BQ, O'Hara R, Kovacevic S, Schulman A, Padovan N, Marinkovic K (2014) Oscillatory spatial profile of alcohol's effects on the resting state: anatomically-constrained MEG. Alcohol 48(2):89–97

Shankar S, Udupi VR (2014) Support vector machine based face recognition. Int J Adv Res Sci Eng 3(1):119–126

Ye QL, Ye N, Zhang HF, Zhao CX (2014) Fast orthogonal linear discriminant analysis with applications to image classification. In: International joint conference on neural networks (IJCNN)

Yeddula SR (2012) Applications of hilbert huang transform (HHT) to the analysis of EEG and other non-stationary time series. Lamar University

Zong C, Chetouani M (2009) Hilbert-huang transform based physiological signals analysis for emotion recognition. Institut des Systèmes Intelligents et de Robotique 1–6

Biosurfactant Production by *Pseudomonas fluorescens* NCIM 2100 Forming Stable Oil-in-Water Emulsions

Neha Panjiar, Shashwati Ghosh Sachan and Ashish Sachan

Abstract

Pseudomonas fluorescens NCIM 2100 was studied for its potential to produce extracellular biosurfactant in the nutrient broth medium supplemented with diesel (2%) as an additional carbon source. Biosurfactant production was checked based on oil spread and drop collapse test. Emulsification index test was conducted to determine the emulsifying ability of the crude biosurfactant produced. The emulsification index, minimum surface tension and production of crude bioemulsifier were found to be 90.38% with diesel, 34.40 ± 0.03 mN/m and 2.80 ± 0.45 g/L respectively. Maximum product accumulation occurred during stationary phase of the bacterial growth curve. Crude biosurfactant produced was stable, withstanding a wide temperature (37–80 °C) and pH range (7–10.5), with an E-24 Index value greater than 50%. Produced biosurfactant has an efficient emulsifying ability. Most of the emulsions formed with tested aliphatic hydrocarbons and ester-based vegetable oils were stable. Emulsions formed with different hydrophobic substrates were oil-in-water (o/w) in nature.

Keywords

Pseudomonas fluorescens NCIM 2100 · Biosurfactant · Oil-in-water emulsion · Droplet size measurement

N. Panjiar · S.G. Sachan · A. Sachan (✉)
Department of Bio-Engineering, Birla Institute of Technology, Mesra, Ranchi 835215, Jharkhand, India
e-mail: asachan@bitmesra.ac.in

S.G. Sachan
e-mail: ssachan@bitmesra.ac.in

K. Mukhopadhyay et al. (eds.), *Applications of Biotechnology for Sustainable Development*, DOI 10.1007/978-981-10-5538-6_13

Introduction

Biosurfactants are surface active compounds of microbial origin (Panjiar et al. 2015). Biodegradable nature, low toxicity, chemical and functional diversity makes these compounds a suitable substitute for chemical surfactants (Panjiar et al. 2013). Presence of both hydrophilic and hydrophobic moieties causes its accumulation at the air–water or oil–water interface, thereby reducing surface and interfacial tension respectively (Vijayakumar and Saravanan 2015). Bioemulsifiers are group of biosurfactants having efficient emulsification properties, generally comprising of high molecular weight biosurfactants (Lovaglio et al. 2015). They form microemulsions where hydrophobic substrates can disperse in aqueous phase, forming oil-in-water (o/w) emulsions, or water can disperse in oil as continuous phase resulting into water-in-oil emulsions (w/o). Oil-in-water emulsions having characteristic aqueous continuous phase have found applications in different industries like cosmetics, food, pharmaceutical, agriculture and petroleum (Randhawa and Rahman 2014). Industrial process like microbial enhanced oil recovery (MEOR) requires the use of thermostable biosurfactants (Pacwa-Płociniczak et al. 2011). The genus *Pseudomonas* has been extensively studied for its biosurfactant producing potential (Aparna et al. 2012; Rebello et al. 2013). Although the research related to large-scale biosurfactant production has considerably increased during recent years, target biotechnological production of these compounds at industrial scale is yet to achieve, which is the requirement of the present time (Rebello et al. 2013). The purpose of the present work is to determine the biosurfactant production capability of *Pseudomonas fluorescens* NCIM 2100 and its application in formation of stable oil-in-water emulsions.

Materials and Methods

Maintenance of the Microorganism and Preparation of Seed Culture

The bacterial culture, *P. fluorescens* NCIM 2100 was obtained from the culture collection of Department of Bio-Engineering, Birla Institute of Mesra, Ranchi, Jharkhand, India, and was regularly subcultured after an interval of one month and maintained on nutrient agar slants at 4 °C for future use. Preparation of seed culture was performed by transferring a loopfull of microorganism from the stored slant into 6 mL of sterilized nutrient broth followed by incubation at 37 °C and 120 rpm in a shaker incubator (New Brunswick Scientific, USA) for 12–15 h.

Detection of Biosurfactant in the Culture Broth

Detection of biosurfactant in the culture broth of *P. fluorescens* NCIM 2100 was determined by inoculating 1% (v/v) of homogenous seed culture ($OD_{600} \sim 1$) in nutrient broth medium supplemented with diesel (2%) followed by incubation at 37 °C and 120 rpm. Hydrophobic substrate, diesel, was used to induce the production of biosurfactant. Culture supernatant was obtained after centrifugation at 5000 rpm for 20 min at 25 °C. Biosurfactant production in the culture supernatant was checked using multiple screening tests like oil spread (Morikawa et al. 2000), drop collapse (Jain et al. 1991) and emulsification index (E-24 index) test (Cooper and Goldenberg 1987), as described earlier by Panjiar et al. (2015). The percentage of emulsification index is calculated by using Eq. (1). Sodium dodecyl sulphate (SDS) and phosphate buffer saline (PBS) were used as positive and negative controls respectively.

$$\text{E-24 Index, } \% = (\text{Height of emulsion formed}/\text{Total height of solution}) \times 100 \quad (1)$$

Surface tension measurement of the culture supernatant and whole culture broth was performed by the dynamic contact angle tensiometer DCAT21 (Dataphysics, Germany) using the Wilhelmy plate method. Sterile nutrient broth was used as control (Nitschke et al. 2004).

Bacterial Growth and Biosurfactant Production Study

The growth pattern of *P. fluorescens* NCIM 2100 was determined both in the presence and absence of diesel as described earlier by Panjiar et al. (2015). Microbial growth (OD at 600 nm) and biosurfactant production, measured in terms of E-24 Index, was determined simultaneously by withdrawing samples from the culture broth.

Biosurfactant Extraction from the Culture Broth

Biosurfactant produced by *P. fluorescens* NCIM 2100 into the broth culture supernatant was acid hydrolyzed with concentrated hydrochloric acid to precipitate it, followed by overnight incubation at 4 °C. During acidification, biosurfactant present in protonated form become less soluble in aqueous solution. The precipitated bioemulsifier was collected by centrifugation at 5000 rpm for 20 min at 4 °C. Collected bioemulsifier was washed twice with distilled water, dried at 65 °C and weighed to calculate its production (Nitschke et al. 2004).

Stability Study of Extracellular Crude Biosurfactant

Effect of pH and Temperature

Stability of the extracellular crude biosurfactant was determined by measuring the E-24 Index of the culture supernatant after subjecting it to different pH and temperature ranges. pH values of 2–10 was adjusted by hydrochloric acid (1N)/sodium hydroxide (1N). Temperatures considered for study were 4, 10, 20, 40, 60 and 80 °C (Khopade et al. 2012).

Emulsifying and Emulsion Stabilizing Capacity of the Extracellular Biosurfactant

Emulsifying ability of the produced biosurfactant was examined by studying E-24 Index with different hydrophobic substrates (hexane, heptane, octane, hexadecane, dodecane, benzene, toluene, xylene, diesel, kerosene and oils of soybean, mustard, groundnut, olive and coconut). The emulsion formed by vortexing hydrophobic substrates and culture supernatant was left to stand for 1 h, and was considered as starting time, 0 h. Emulsification Index (E-24 Index, %), relative emulsion volume (EV, %), emulsion stability (ES, %) and emulsified organic phase (EOP, %) were calculated at 24 h intervals up to 72 h, 1 month and then after 3 months' intervals from Eqs. (1) to (4) respectively (Batista et al. 2006; Portilla-Rivera et al. 2010).

$$\text{EV, \%} = \left(\text{emulsion height, cm} \times \text{cross section area, cm}^2\right) \times 100/\text{total liquid volume, cm}^3 \tag{2}$$

$$\text{ES, \%} = (\text{EV at time h} \times 100)/\text{EV at 0 h} \tag{3}$$

$$\text{EOP, \%} = \left[\text{TOP, cm}^3 - \left(\text{NEOP, cm} \times \text{cross section area, cm}^2\right) \times 100\right]/\text{vol TOP, cm}^3 \tag{4}$$

where TOP is total volume of organic phase and NEOP is non-emulsified organic phase. The emulsions stabilized by culture supernatant were also compared with those formed by 1% (w/v) solution of the chemical surfactant SDS in deionized water.

Nature and Droplet Size Distribution Study of the Emulsions Formed

Nature of the emulsion, whether it was o/w or w/o, was determined by the addition of small amount of powder water soluble dye (methylene blue) on the surface of the tested emulsion kept on the slide. If the external continuous phase was aqueous (emulsion of o/w type), the dropped dye got diffused immediately throughout the aqueous phase of the emulsion. On contrary, if the emulsion was of w/o type, the dropped dye remained as clump on the surface of the tested emulsion (http://www.surfatech.com/pdfs/emulsions.pdf).

Optical microscope was used to measure the size of the emulsion's droplets (Portilla-Rivera et al. 2010). A drop of emulsion was placed in a glass slide and was observed through a 10× objective lens of the microscope. Radii of the observed droplets were measured through standard micrometric procedure. Images of several regions from each slide were taken to capture the representative structure of emulsion droplets.

Results and Discussion

P. fluorescens NCIM 2100 was investigated to produce surface active compounds called biosurfactants.

Detection of Biosurfactant in the Culture Broth

Sensitive detection methods like the drop collapse and oil spread tests are included together for primary screening of biosurfactant production (Satpute et al. 2008), whereas E-24 index test was conducted to determine the emulsifying ability of the produced biosurfactant (Dhail and Jasuja 2012). The drop containing culture supernatant collapsed within 10 s, thereby indicating the presence of biosurfactant. Diameter of clear zone was observed to be greater than 5 cm, further confirming the surfactant nature of the culture supernatant. Maximum E-24 Index observed was 90.38% with diesel, which is in conformity with Willumsen and Karlson (1997), stating a promising bioemulsifier should have an E-24 Index greater than 50%. Biosurfactant produced by *P. fluorescens* NCIM 2100 could reduce the surface tension value of the culture broth to 34.40 ± 0.03 mN/m, which was around the threshold value of 40 mN/m or lower (Olivera et al. 2003). Significant reduction of surface tension was not observed may be due to production of polymeric biosurfactant having reasonable emulsification abilities as compared to surface activity (Willumsen and Karlson 1997; Plaza et al. 2006). Extracellular nature of the produced biosurfactant was confirmed by the similar surface tension values of culture broth and culture supernatant.

Bacterial Growth and Biosurfactant Production Study

Simultaneous study was conducted for measuring bacterial growth (A_{600}) and biosurfactant production (E-24) at different time intervals (Fig. 1). Biosurfactant was not detected in the nutrient broth without diesel, which is control sample. Considerable bioemulsifier production started after 32 h of incubation period at the

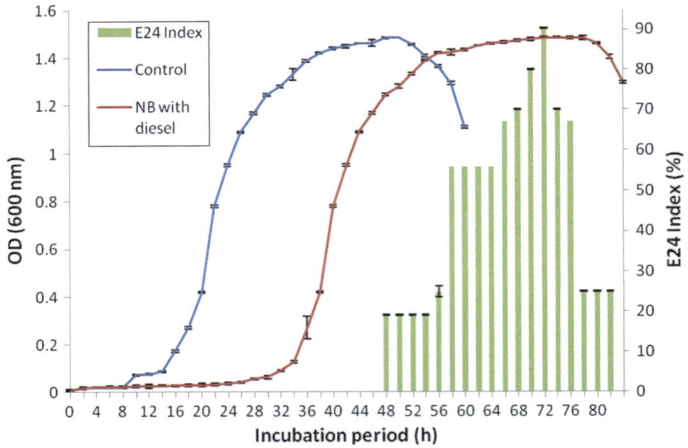

Fig. 1 Growth kinetics and biosurfactant production profile in terms of E-24 Index (%) of *P. fluorescens* NCIM 2100. *Error bars* indicate standard deviation

exponential phase of bacterial growth curve, when grown in production medium. Highest production (E-24 Index of 90.38%) was attained after 72 h of incubation, at its stationary phase of nutrient limiting condition. Singh and Tripathi (2013) have described maximum production of rhamnolipid biosurfactant during stationary phase of the studied bacterial strain *Pseudomonas stutzeri*. Different authors have also reported biosurfactant production during their stationary phase (Batista et al. 2006; Sriram et al. 2011). 2.80 ± 0.45 g/L of biosurfactant was produced, as determined by acid precipitation method. The precipitate was again dissolved in distilled water and pH was set to 7 by adding sodium hydroxide (1N) and it was checked for the presence of bioemulsifier by emulsification index test. The remaining culture supernatant was also checked and found to give negative result for emulsification index test.

Stability Study of Extracellular Crude Biosurfactant

Effect of pH and Temperature

Crude bioemulsifier produced by *P. fluorescens* NCIM 2100 is stable at pH range of 7–10.5 (Fig. 2a) and temperature range of 37–80 °C (Fig. 2b) with E-24 Index value greater than 50%. Thermal and pH stability of the crude biosurfactant render its applicability in process like MEOR (Singh and Tripathi 2013).

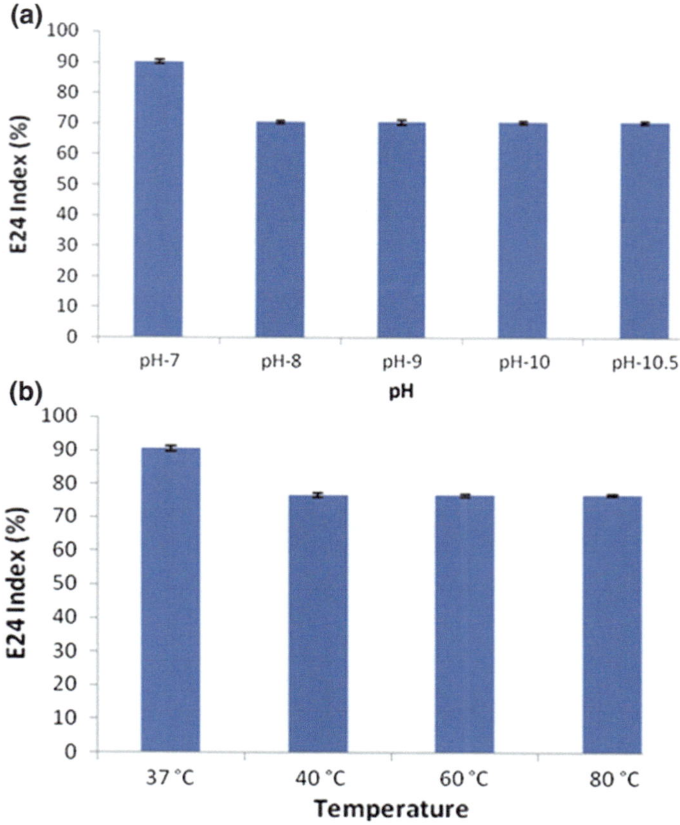

Fig. 2 Stability of the crude biosurfactant at different. **a** pH and **b** temperature. *Error bars* indicate standard deviation

Emulsifying and Emulsion Stabilizing Capacity of the Extracellular Biosurfactant

It was evaluated by determining E-24 Index (%), EV (%), ES (%) and EOP (%) with different hydrophobic substrates namely hexane, heptane, octane, hexadecane, dodecane, benzene, toluene, xylene, diesel, kerosene, soybean, mustard, groundnut, olive and coconut oil at 0, 24 h, 1 month and 3 months' intervals. Biosurfactant produced by *P. fluorescens* NCIM 2100 could efficiently emulsify all the tested aliphatic and ester-based oils but was less capable of emulsifying aromatic hydrocarbons. From Fig. 3a–d, it was apparent that emulsions formed with diesel and other aliphatic hydrocarbons had a greater E-24 index (90.38%), EV values (75.98%) after olive oil (98.07% E-24 Index and 97.97% EV). All the emulsions formed were 100% stable up to 24 h, with the entire organic layer converted into emulsion. However, less emulsion stability was detected in case of mustard and coconut oil.

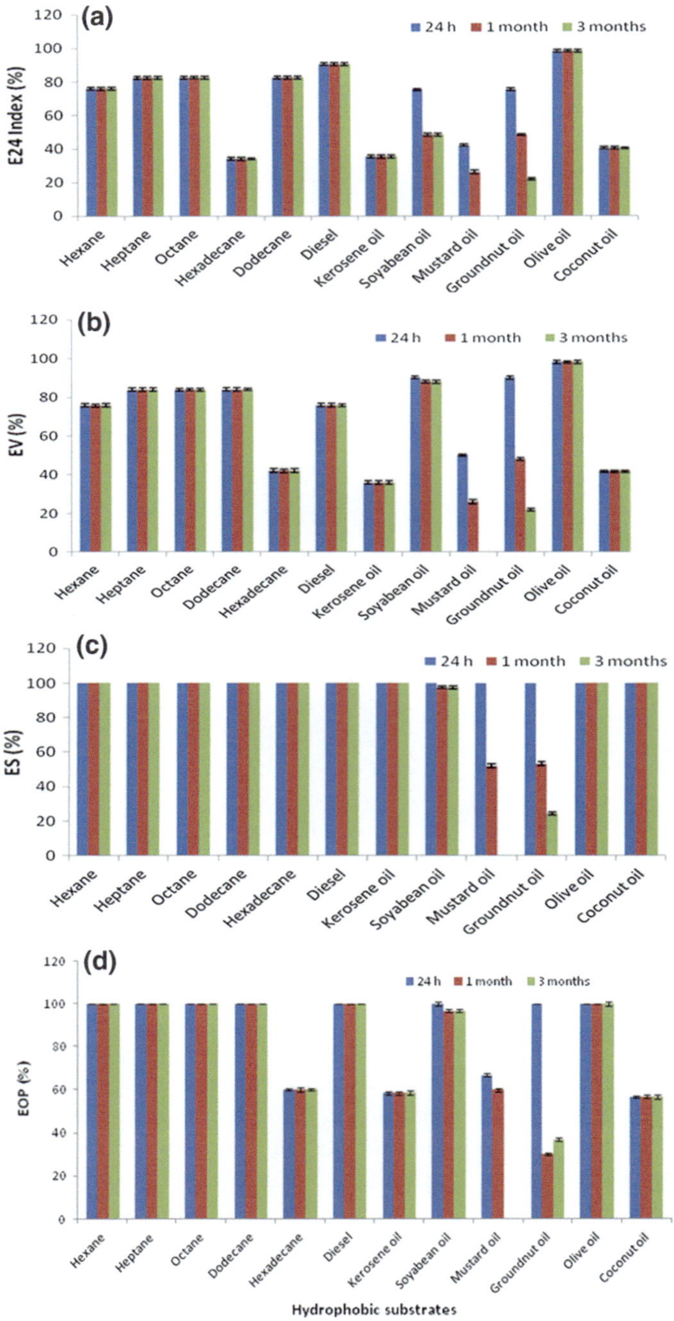

Fig. 3 E-24 Index, EV, ES and EOP (**a**, **b**, **c** and **d** respectively) of emulsions produced by crude biosurfactant with different hydrophobic substrates at 24 h, 1 and 3 months interval. *Error bars* indicate standard deviation

Emulsifying ability and stability results of biosurfactant by studied bacteria were found to be comparable to the chemical surfactant used (SDS) with 75% of E-24 Index, EV and 100% of ES and EOP. PBS as negative control was found to have 0% of E-24 Index, EV, ES and EOP.

Study by Aparna et al. (2012) has reported different E-24 Index values with sunflower oil, crude oil, gasoline, *n*-hexadecane, kerosene, hexane and benzene by the isolate *Pseudomonas* sp. 2B, when grown in modified proteose peptone glucose ammonium salt medium. Maximum E-24 Index value of 84% was found with sunflower oil, while the produced biosurfactant was not able to emulsify benzene. The results were found to be similar as in the present study.

Nature and Droplet Size Distribution Study of the Emulsions Formed

Emulsions formed with different hydrophobic substrates by the produced biosurfactant, was found to be o/w in nature. Optical images of emulsion apparently showed the oil droplets dispersed in continuous aqueous phase (Fig. 4b). Figure 4c showed the compact o/w emulsion of olive oil in water.

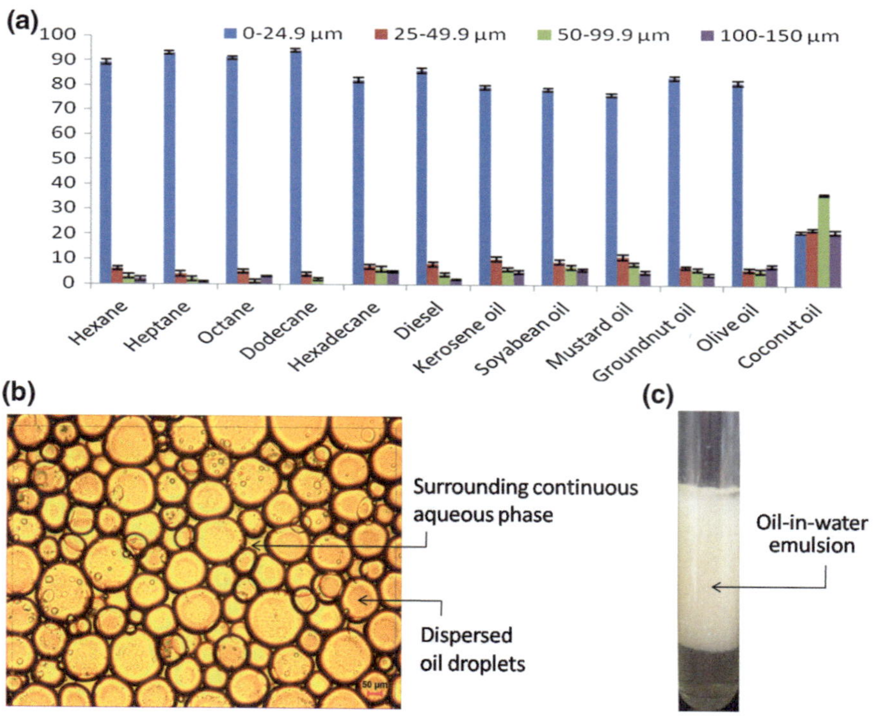

Fig. 4 **a** Radius distribution of droplets for the emulsions stabilized with the crude biosurfactant. **b** Optical image of the o/w emulsion. **c** Emulsion of olive oil in water

Standard micrometric procedure was used to measure the droplet size of the emulsions. Figure 4a represents the distribution of droplets diameter for different emulsions. It was observed that compact and stable emulsions comprised of more than 75% of droplets of diameter between 0 and 24.9 μm. Whereas loose and unstable emulsions contain 5.5–36 and 7.5–21% of droplets of diameter 50–99.9 and 100–150 μm respectively. Moreover, it was also observed that compact emulsions were polydispersed in nature, characterized by droplets of different diameter whereas loose emulsions are more homogeneous with larger droplet size, which is in accordance with the findings of Portilla-Rivera et al. (2010).

Conclusion

P. fluorescens NCIM 2100 was examined for its biosurfactant-producing property in the nutrient-rich medium containing diesel as an additional carbon source. Stationary phase of the bacterial growth was observed to be the most optimal for maximum biosurfactant accumulation into the medium. Stability of the crude biosurfactant at broad pH and temperature range and its ability to efficiently emulsify aliphatic hydrocarbons and ester-based vegetable oils, forming stable oil-in-water emulsions, suggests its potential use in petroleum, food, cosmetic industry and as cleaning agents. It may have environmental applications for mineralization of recalcitrant hydrocarbons in contaminated areas. Moreover, the use of obtained biosurfactant in its crude form conform the requirement of cost reduction.

Acknowledgements Neha Panjiar acknowledges Council of Scientific and Industrial Research (CSIR), New Delhi for the research fellowship provided [09/554(0023)/2010-EMR-I] and corresponding authors acknowledge the University Grant Commission (UGC), Government of India for the financial support [F. No. 40-160/2011(SR)]. The authors are grateful to Central Instrumentation Facility (CIF) and Department of Bio-Engineering at Birla Institute of Technology, Mesra, Ranchi for providing culture and instrumentation facilities necessary to carry out the work.

References

Aparna A, Srinikethana G, Smitha H (2012) Production and characterization of biosurfactant produced by a novel *Pseudomonas* sp. 2B. Colloids Surf B 95:23–29

Batista SB, Mounteer AH, Amorim FR, Tótola MR (2006) Isolation and characterization of biosurfactant/bioemulsifier-producing bacteria from petroleum contaminated sites. Bioresour Technol 97:868–875

Cooper DG, Goldenberg B (1987) Surface active agents from two *Bacillus* species. Appl Environ Microbiol 53:224–229

Dhail S, Jasuja ND (2012) Isolation of biosurfactant-producing marine bacteria. Afr J Environ Sci Technol 6:263–266

Jain DK, Thompson DKC, Lee H, Trevors JT (1991) A drop collapsing test for screening surfactant producing microorganisms. J Microbiol Methods 13:271–279

Khopade A, Biao R, Liu X, Mahadik K, Zhang L, Kokare C (2012) Production and stability studies of the biosurfactant isolated from marine *Nocardiopsis* sp. B4. Desalination 285: 198–204

Lovaglio RB, Silva VL, Ferreira H, Hausmann R, Contiero J (2015) Rhamnolipids know-how: looking for strategies for its industrial dissemination. Biotechnol Adv. doi:10.1016/j.biotechadv.2015.09.002

Morikawa M, Hirata Y, Imanaka T (2000) A study on the structure function relationship of the lipopeptide biosurfactants. Biochem Biophy Acta 1488:211–218

Nitschke M, Ferraz C, Pastore GM (2004) Selection of microorganisms for biosurfactant production using agroindustrial wastes. Braz J Microbiol 35:81–85

Olivera NL, Commendatore MG, Delgado O, Esteves JL (2003) Microbial characterization and hydrocarbon biodegradation potential of natural bilge waste microflora. J Ind Microbiol Biotechnol 30:542–548

Pacwa-Płociniczak M, Płaza GA, Piotrowska-Seget ZZ, Cameotra SS (2011) Environmental applications of biosurfactants: recent advances. Int J Mol Sci 12:633–654

Panjiar N, Gabrani R, Sarethy IP (2013) Diversity of biosurfactant producing *Streptomyces* isolates from hydrocarbon-contaminated soil. Int J Pharm Bio Sci 4:524–535

Panjiar N, Ghosh Sachan S, Sachan A (2015) Screening of bioemulsifier-producing micro-organisms isolated from oil-contaminated sites. Ann Microbiol 65:753–764

Plaza GA, Zjawiony I, Banat IM (2006) Use of different methods for detection of thermophilic biosurfactant producing bacteria from hydrocarbon-contaminated and bioremediated soils. J Petrol Sci Eng 50:71–77

Portilla-Rivera OM, Torrado AM, Domínguez JM, Moldes AB (2010) Stabilization of kerosene/water emulsions using bioemulsifiers obtained by fermentation of hemicellulosic sugars with *Lactobacillus pentosus*. J Agric Food Chem 58:10162–10168

Randhawa KKS, Rahman PKSM (2014) Rhamnolipid biosurfactants-past, present, and future scenario of global market. Front Microbiol 5:454–460

Rebello S, Asok AK, Joseph SV, Joseph BV, Jose L, Mundayoor S, Jisha MS (2013) Bioconversion of sodium dodecyl sulphate to rhamnolipid by *Pseudomonas aeruginosa*: a novel and cost-effective production strategy. Appl Biochem Biotechnol 169:418–430

Satpute SK, Bhawsar BD, Dhakephalkar PK, Chopade BA (2008) Assessment of different screeningmethods for selecting biosurfactant producing marine bacteria. Indian J Mar Sci 37:243–250

Singh DN, Tripathi AK (2013) Coal induced production of a rhamnolipid biosurfactant by *Pseudomonas stutzeri*, isolated from the formation water of Jharia coalbed. Bioresour Technol 128:215–221

Sriram MI, Kalishwaralal K, Deepak V, Gracerosepat R, Srisakthi K, Gurunathan S (2011) Biofilm inhibition and antimicrobial action of lipopeptide biosurfactant produced by heavy metal tolerant strain *Bacillus cereus* NK1. Colloids Surf B 85:174–181

Vijayakumar S, Saravanan V (2015) Biosurfactants-types, sources and applications. Res J Microbiol 10:181–192

Willumsen PA, Karlson U (1997) Screening of bacteria isolated from PAH-contaminated soils for production of biosurfactants and bioemulsifiers. Biodegradation 7:415–423

Identification and Screening of Potent Inhibitors Against Spore Wall Proteins of Flacherie Infected *Bombyx mori* Through Molecular Modeling and Docking Studies

Debadyuti Banerjee and Koel Mukherjee

Abstract

Sericulture or cultivation of silkworms is practiced in India and other Asian countries since time immemorial. The mulberry silkworm, *Bombyx mori* are affected by viral, fungal and protozoan pathogens, which causes huge monetary loss. Among many other diseases, Flacherie is the most rampant in the silkworm community causing flaccidity and subsequent death. Previous studies indicate *Andrographolide*, *Quercitin* and chitosan are the active compounds from the extracts of *Andrographis paniculata* and *Tridax procumbens* to be effective against the flacherie pathogens. The main aim of the following work depends on the screening and selecting of the potent inhibitor against the spore wall proteins of flacherie-infected *B. mori*. In this study, two wall proteins were selected from flacherie-infected mulberry silkworm, the 3D structures were modeled through Modeller 9.15. Eventually the modeled structures were docked with reference inhibitors (three best reference) using GLIDE suite of Maestro 9.3.5. The best one among the three references was identified and selected for the screening of next set of inhibitors (50 inhibitors) with a structural similarity of 70, 80 and 90%. They were again docked in the same active region of the targets and identified as the best reference inhibitor. The result shows a promising potent inhibitor which can be further validated by experimental procedures.

Keywords

B. mori · Flacherie · Spore wall protein · GLIDE

D. Banerjee · K. Mukherjee (✉)
Department of Bio-Engineering, Birla Institute of Technology,
Mesra, Ranchi 835215, Jharkhand, India
e-mail: koelmukherjee@bitmesra.ac.in

K. Mukhopadhyay et al. (eds.), *Applications of Biotechnology for Sustainable Development*, DOI 10.1007/978-981-10-5538-6_14

Introduction

Cultivation of silkworm for the production of fiber is called as "Sericulture". Silk is considered to be "Queen of Fiber" which is proteinaceous in nature. Archeological and bibliographical evidences show that the sericulture was practiced in China about 2500 BC (Wood 2002). In twelfth century BC, it expanded outside China as mulberry seed and silkworm egg were smuggled out. During 1920–30s world silk yield exceeded 60,000–70,000 tons, while in 1950–60s, it decreased to 30,000 tons. After 1970, showed rapid development and had become the vocation of small agricultural families in popular developing countries like, China, India, Vietnam and Thailand (Patil et al. 2009). Current annual production of silk is about 496,000 tons. India is the second largest producer of silk and the fact that nearly 6 million Indian's are involved in sericulture (http://www.nistads.res.in/indiasnt2008/t6rural/t6rur16.htm).

Silkworm diseases are the most important disease that inflects heavy loss to crops. The losses occur mainly during the final stages of silkworm rearing resulting considerable energy and money loss. The mulberry silkworm, *B. mori* known for producing silk cocoons are affected by viral, fungal, and protozoan pathogens among which bacterial pathogens independently cause cocoon loss to the tune of 75% (Das Gupta 1950; Rajitha and Savithri 2015). Bacteria such as *S. faccalis*, *S. liquifacions*, *S. acire*, *S. epidermidis*, and *Bacillus* sp. are commonly reported to cause flacherie in silkworm (Adhithya et al. 2013; Anitha et al. 1994). Flacherie is a flaccid disease common in silkworms (Manimegalai and Chandramohan 2005). The dead larvae, gut juice, feces, and alternate hosts serve as the sources of infection. Flacherie is an important disease affecting silkworm, which is caused by microbial agents (Sakthivel et al. 2012). During the initial stages of infection, the larva stops feeding and show sluggish movement. Fecal matter of diseased or dead silkworms, parallel to the contaminated mulberry leaves act as a good source of infection. Bad rearing, fluctuation in temperature, and high humidity are also the major influencing factors for flacherie disease (Nataraju et al. 1999). The leaves of poor nutritive value will not be able to provide sufficient quality of essential requirement to the larva to produce antibacterial factor, which results in high rate of multiplication of infectious bacteria and development of bacterial flacherie. Selvakumar et al. (2002) estimated the crop loss of about 27–35% with 11–15 kg cocoon loss/100 dfls.

However, when silkworms are physiologically weak, bacterial diseases can attack them, eliciting a heavy toll on sericulture. Presently there are no potent inhibitors against the disease, but scientists are always provoked to search for novel sources (Saxena and Sosanna 2005) of biocides with broad-spectrum activities (Abad et al. 2007). Some sources report extracts of *Tridax* (Bhringraja) and *Andrographis* (Kalmegh) as effective against the pathogens (Karthikairaj et al. 2013).

Silk productivity could be improved by better understanding the disease, its pathway and designing of suitable inhibitors against the spore wall proteins, thereby reducing the incidence of silkworm diseases. Nowadays there are no NMR or X-ray crystallographic structures of the wall protein of *B. mori.* So, in this study first we are conducting an in silico structure-based design of potent inhibitors against the spore wall protein of flacherie-infected silkworms. Two 3D structures of cuticular proteins were homology modeled and docked with the reference ligands. Andrographolide, Quercitin, and chitosan are taken as reference inhibitors. The designed inhibitors could be synthesized and brought into the market for effective control of this disease which is rampant during silkworm rearing which causes considerable energy and money.

Materials and Methods

Identification and Selection of Wall Proteins

The amino acid sequences of two wall proteins, with accession I.D XP_004934014.1 and Q09GR7 were retrieved from NCBI (www.ncbi.nlm.nih.org) and Uniprot databases (http://www.uniprot.org) respectively.

Model Building and Validation

The wall proteins were modeled using Modeller 9.15 by taking templates from HHPRED (http://toolkit.tuebingen.mpg.de/hhpred). The models with the lowest DOPE (Discrete Optimized Protein Energy) were taken into consideration. The Ramachandran plot of the models was obtained from the RAMPAGE server (http://mordred.bioc.cam.ac.uk/~rapper/rampage.php). The models were also verified using SAVES (Structural Analysis and Verification Server). The stereochemical quality of the model was evaluated with PROCHECK (Laskowski et al. 1993) by Ramachandran plot analysis. It was further analyzed by ERRAT (Colovos and Yeates 1993) and VERIFY3D (Bowie et al. 1991).

Molecular Dynamics Simulation

From a dynamic point of view, the MD Simulation was performed with GROMOS96 43al protein force fields within the periodic boundary conditions (PBC) by using GROMACS 4.6.5 package (Van Der Spoel et al. 2005). The 3D structure of the protein was taken in a cubic box with a 1.0 Å edge length in case of first model and 0.5 Å in case of second. To solvate the condition, the "SPC" water model (spc216.gro file) was used to fill up the box. Before simulation, energy minimization was performed by steep and conjugate gradient (cg) methods to examine whether the predicted structure of protein remains stable or not. MD simulation was performed for 10 ns maintaining the temperature at 300 K and pressure at 1 atm.

After completion of simulation, different trajectory files were generated which were analyzed by different tools of GROMACS.

Ligand Binding Site Prediction

The active sites of the predicted structures were searched by using COACH server (http://zhanglab.ccmb.med.umich.edu/COACH/). COACH is a meta-server approach to protein-ligand binding site prediction (Sastry et al. 2013). The active sites of wall protein 1 are ASP-116, ASP-119 and ASP-134, whereas the active sites of wall protein 2 are VAL-6, VAL-7, LEU-9, ALA-10, LEU-11, VAL-12, ALA-13, PRO-107, GLN-108 respectively.

Ligand Preparation and Docking

Docking is a method which predicts the preferred orientation of one molecule to a second when bound to each other to form a stable complex. The docking was done by using Glide application of Maestro 9.3.5 (Friesner et al. 2004). The structure of ligands, *Andrographolide*, *Quercitin* and chitosan were downloaded from Zinc database (www.zinc.docking.org). The selected inhibitors were prepared using LigPrep module of Maestro. Grid for the binding site was prepared using the coordinates of active site residues of the wall proteins. GLIDE docking was started after using the grid the ligand preparation. Extra precision (XP) docking was performed in this study.

Results and Discussion

Model Building

The 3D structure of protein models (wall proteins) was constructed using Modeller 9.15 (https://salilab.org/modeller/). For each of the cases, five models were generated. The models were selected based on the lowest DOPE score. The DOPE scores of the selected models (SEQ1B99990002 and SEQ5B99990005) were −3870.97241 and −6197.13672 respectively. The selected 3D structures of the predicted models were represented in Fig. 1.

Model Validation

The modeled structures were analyzed by using Rampage and SAVES servers. The number of residues in favored regions is 92.7% in case of first model (SEQ1B99990002) and 90.2% in case of second model (SEQ5B99990005) as

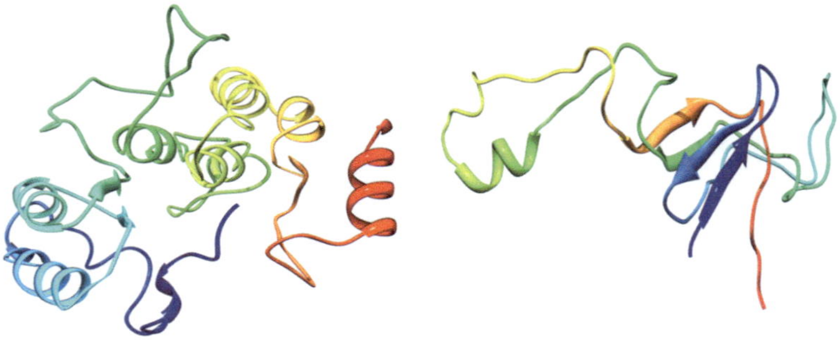

Fig. 1 Ribbon representation of modeled structures through modeling with Modeller 9.15 of two wall proteins **a** SEQ1B99990002 and **b** SEQ5B99990005

predicted by Rampage. As the selected models are having the highest percentage value of residues in favored region so the models, SEQ1B99990002 and SEQ5B99990005 were selected for the next analysis steps (Rungta et al. 2014). The results of Rampage are represented in Fig. 2. Verify3D results of protein (a) SEQ1B99990002 indicate 87.02% of the residues had an averaged 3D-1D score ≥ 0.2 and that of (b) SEQ5B99990005 indicate 33.33%. The ERRAT overall quality factor of XP_004934014.1 (SEQ1B99990002) and Q09GR7 (SEQ5B99990005) are 37.500 and 21.569. The overall quality factor of ERRAT is expressed as percentage of error values (Colovos and Yeates 1993).

Model Simulation

Molecular dynamics simulation was conducted in GROMACS package 4.6.5. GROMOS96 43al force field is used for simulation. Prior to simulation, energy minimization steps were performed. Both steepest descent and conjugate gradient methods were performed. RMSD and Radius of gyration of atoms of both the models was indicated in Figs. 3 and 4 respectively. RMSD values show the flexibility of the entire protein structure at different time intervals during simulation. In the given plot, the values start from 0 nm showing that there is no deviation from the reference protein structure (Rungta et al. 2014). Initially, however, a decrease in the values can be noticed till 2.7 ns. The RMSD values can be seen to stabilize from 7.5 ns within a range of 1.00E+00 nm to 1.20E+00 nm until 10 ns for the first model and within a range of 6.00E+00 nm to 8.00E+00 nm for the second model. Radius of gyration plot shows distance between the atoms and their center of axis. In the given plot, for the second model shows certain fluctuations in the initial period of 3000 ps while, there were fewer fluctuations in the Radius of gyration values for the first model throughout the time period (Mukherjee et al. 2014).

(a)

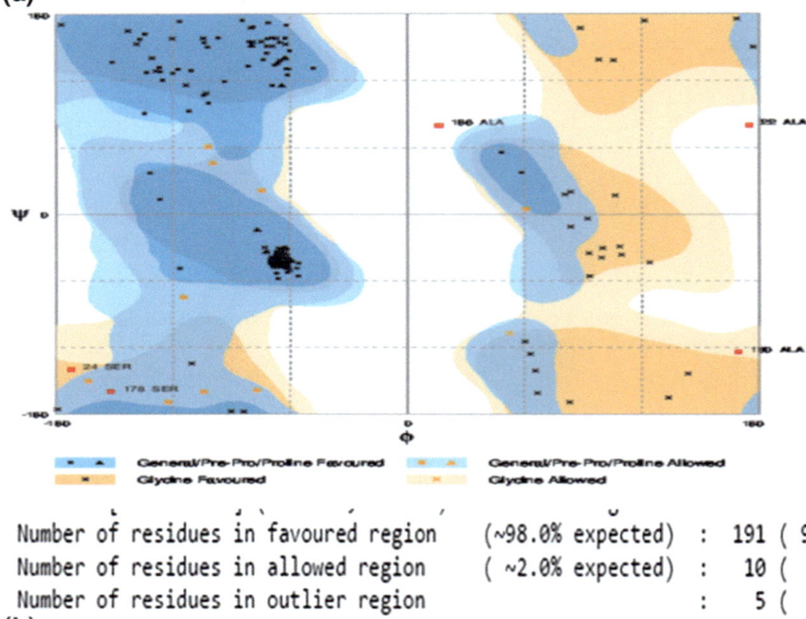

(b)

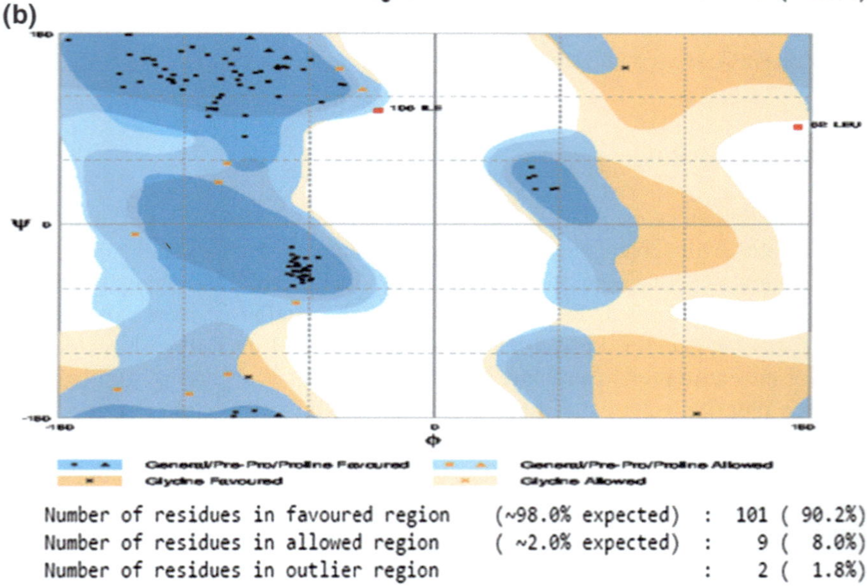

Fig. 2 Ramachandran plots of **a** SEQ1B99990002 and **b** SEQ5B99990005 using RAMPAGE server

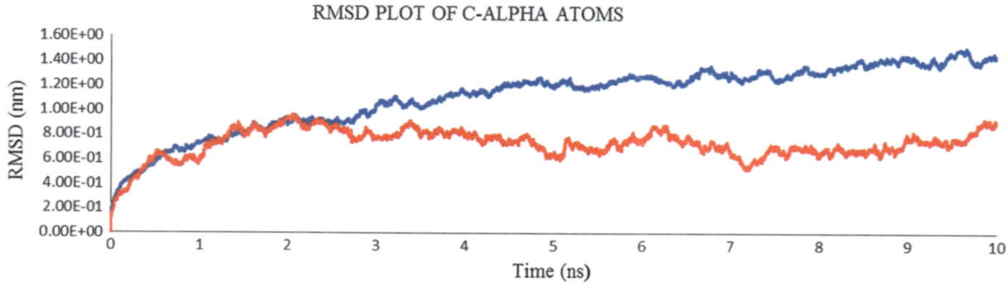

Fig. 3 RMSD plots of both the models, Seq1B9990002 and Seq5B99990005. *Red lines* indicate the second model while *blue lines* indicate the first

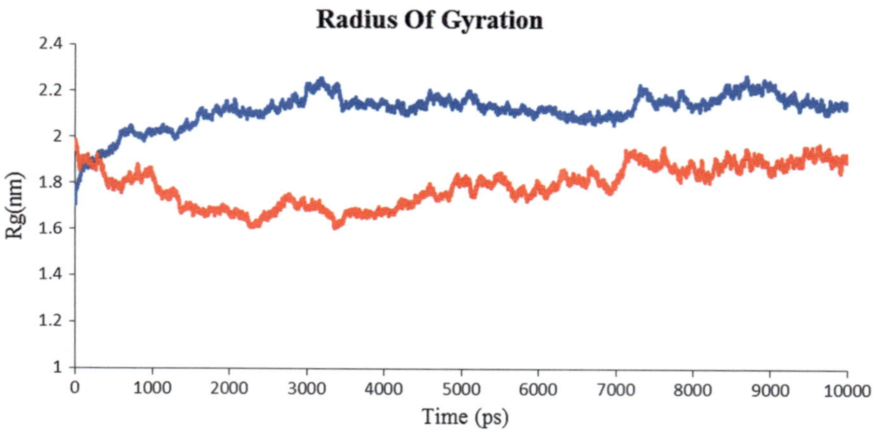

Fig. 4 Radius of gyration plots of both the models, Seq1B9990002 and Seq5B99990005. *Red lines* indicate the second model while *blue lines* indicate the first

Docking Results

Docking of the two proteins with the selected inhibitors were done through GLIDE suite in Maestro 9.3.5. Glide is docking software that offers a wide array of speed and precision from high throughput virtual screening of millions of compounds to extremely accurate binding mode predictions, providing consistently high enrichment at every level (Friesner et al. 2004). Prior to docking, H-bond optimization was performed in ProtAssign module, followed by minimization by Impref module with OPLS 2005 force fields. The inhibitors were prepared using the LigPrep module. Grid was constructed using the coordinates of the active sites of the selected proteins (Maity et al. 2016). Among the three reference inhibitors (Table 1), the complex with least glide score (**−9.002** and **−8.192**) was considered (Roy and Mukherjee 2015) to be the best docked result. Tables 2 and 3 represents the Glide score and docking scores for both the modeled proteins.

Table 1 Properties of selected reference inhibitors

S. No.	Inhibitors	XLogP	Hydrogen bond donors	Hydrogen bond acceptors	Net charge	Molecular weight	Polar desolvation (kcal/mol)
1.	Andrographolide	1.05	3	5	0	350.455	−14.42
2.	Chitosan	−5.09	12	11	2	342.345	−108.2
3.	Quercitin	1.68	5	7	0	302.238	−13.58

Table 2 GLIDE scores and docking scores of docked complexes of model SEQ1B99990002

S. No.	Inhibitor	GLIDE score	Docking score
1.	**Chitosan**	**−9.002**	**−8.643**
2.	Quercitin	−3.438	−3.438
3.	Andrographolide	−1.782	−4.410

Table 3 GLIDE scores and docking scores of docked complexes of model SEQ5B99990005

S. No.	Inhibitor	GLIDE score	Docking score
1.	Quercitin	−7.934	−7.926
2.	**Chitosan**	**−8.192**	**7.833**
3.	Andrographolide	−3.720	−3.720

Structures of docked complexes and their interaction profiles are indicated in Figs. 5 and 6. The above figures indicate the interaction profiles of the docked complexes generated by Maestro 9.3.5 and PLIP servers (https://projects.biotec.tu-dresden.de/plip-web/plip/index).

The red arrows indicate the H-bond interactions and the magenta arrows indicate the H-bond interaction with the side chain. The blue-dotted lines indicated the hydrophobic interactions while the blue lines indicate the H-bond interaction in case of docked complex of midgut protein and *quercitin*.

The above results were selected and undergone 70% similarity search in Zinc database which yielded 50 experimental new inhibitors. In continuation with the previous part of work, the 50 new inhibitors were again docked with the respective proteins to find a potent one. Results were analyzed for docking of SEQ1B99990002 (wall protein 1) and SEQ5B99990005 (wall protein 2) with 50 new selected inhibitors. Before docking in GLIDE suite of Maestro 9.3.5, the ligands were prepared and using LigPrep and Epik of Maestro which also normally does multiple minimizations. The inhibitor (ZINC77312192) with highest binding affinity (Docking score = −12.376166) was selected for further studies. In case of, the inhibitor (ZINC77312198) with the highest binding affinity (Docking score = −11.288666) was selected. The binding affinity is determined based on lowest GLIDE scores generated from GLIDE suite of MASETRO 9.3.5.

Post docking, the protein–ligand interaction plots were generated using Ligplot+ 1.4.5. Tables 4 and 5 shows the GLIDE scores generated after docking with the structurally similar compounds of chitosan. ZINC77312192 and ZINC77312198

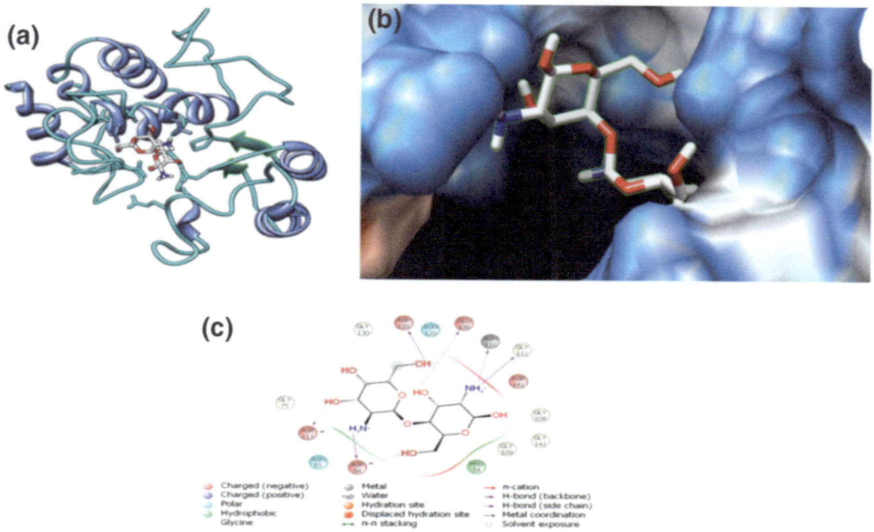

Fig. 5 Docked complex of spore wall protein (SEQ5B99990002) and Chitosan. **a** Ribbon representation of SEQ5B99990002 and chitosan. **b** Close-up view of SEQ5B99990002 and chitosan. **c** 2D interaction profile

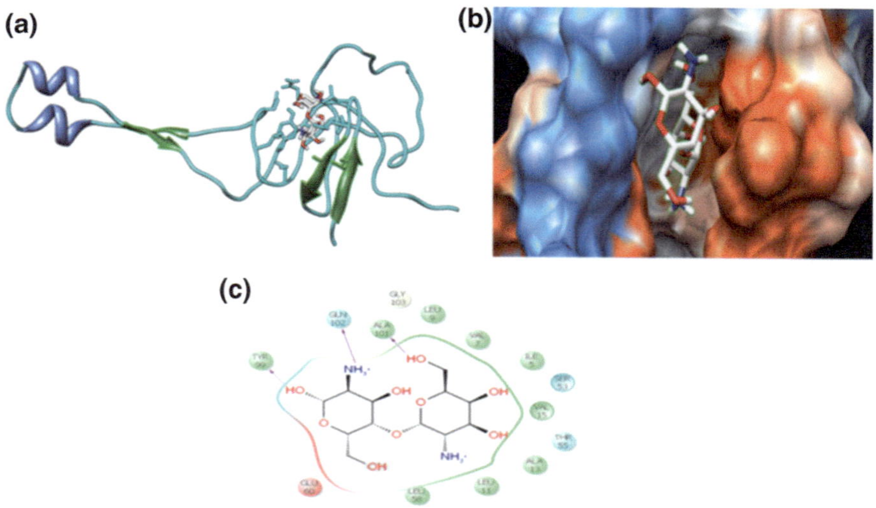

Fig. 6 Docked complex of spore wall protein (SEQ5B99990005) and Chitosan. **a** Ribbon representation of SEQ5B99990005 and chitosan. **b** Close-up view of SEQ5B99990005 and chitosan. **c** 2D interaction profile

Table 4 GLIDE scores of docked complexes of wall protein 1 and first 7 similar compounds of Chitosan among 50 candidates

Ligand ID	GLIDE score
ZINC77312192	**−12.376166**
ZINC71928289	−12.192889
ZINC79114505	−11.788360
ZINC60183167	−11.555782
ZINC77312198	−11.295791
ZINC70665798	−11.150906
ZINC77311627	−11.96849

Table 5 GLIDE scores of docked complexes of Wall protein 2 and first 7 similar compounds of Chitosan among 50 candidates

Ligand Id	GLIDE score
ZINC77312192	−7.451646
ZINC71928289	−8.067042
ZINC79114505	−8.242843
ZINC60183167	−7.491622
ZINC77312198	**−11.288565**
ZINC70665798	−10.096475
ZINC77311627	−9.534169

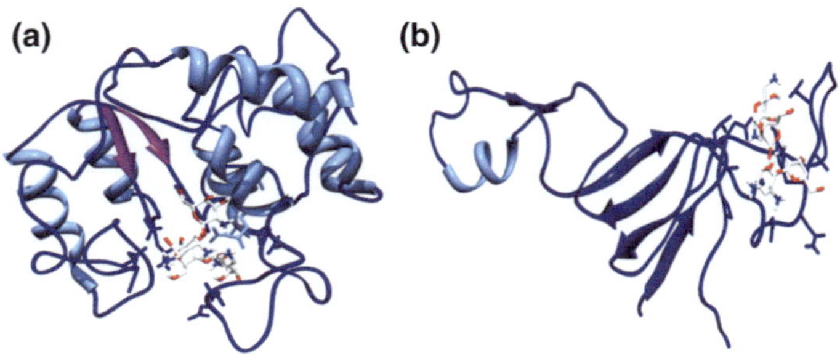

Fig. 7 **a** Ribbon representation of docked complex of wall protein-1 and ZINC77312192. **b** Wall protein-2 and ZINC77312198

shows best binding affinity with wall protein-1 and wall protein-2 respectively (Fig. 7a, b). After docking, the stability of the protein–ligand complex was validated by means of MD simulation for 15 ns

Conclusion

The present works emphasizes the identification, selection of inhibitors against flacherie pathogens with the spore wall proteins and generation of docked complexes with the docking and glide scores. The least GLIDE scores indicate the best affinity of the inhibitor to the wall proteins, which is useful for the identification of the best inhibitor among the group of inhibitors. From the results, chitosan seems to be the best inhibitor among the three inhibitors. Since presently there are no NMR or X-ray crystallographic structures of the wall proteins of *B. mori*, so the modeled structures can be useful for further studies. The selected inhibitor can be taken to the wet lab for its synthesis and further for large scale production. The disease flacherie may be brought under control which would help the sericulture community by bringing the disease under control, thus improving the variety of silk.

Acknowledgements The authors acknowledge BTISnet, Department of Biotechnology, Government of India, New Delhi and TEQIP-II, BIT Mesra for providing Infrastructure facility for the mentioned study. The authors also acknowledge Anand Boipai and Deepak Toppo for providing information for the mentioned study.

References

Abad MJ, Ansuategui M, Bermejo P (2007) Active antifungal substances from natural sources. Arkivoc 2:116–145

Adhithya S, Manimegalai R, Shanmugam R (2013) Morphological and biochemical characterization of bacterial and viral pathogens infecting mulberry silkworm *Bombyx mori L.*. Trends Biosci 6(4):407–411

Anitha T, Sironmani P, Meena P, Vanitha R (1994) Isolation and characterization of pathogenic bacterial species in the silkworm *Bombyx mori L.* Sericologia 34(1):97–102

Bowie JU, Luthy R, Eisenberg D (1991) A method to identify protein sequences that fold into a known three-dimensional structure. Science 253(5016):164–170

Colovos C, Yeates TO (1993) Verification of protein structures: patterns of nonbonded atomic interactions. Protein Sci 2(9):1511–1519

Das Gupta MR (1950) Diseases of silkworm monograph on cottage industries. No. 1, Govt. India Press, Calcutta, p 25

Friesner RA, Banks JL, Murphy RB, Halgren TA, Klicic JJ, Daniel T, Mainz RMP, Knoll EH, Shelley M, Perry JK, Shaw DE, Francis P, Shenkin PS (2004) GLIDE: a new approach for rapid, accurate docking and scoring. 1. Method and assessment of docking accuracy. J Med Chem 47:1739–1749

Karthikairaj K, Prasanna Kumar K, Isaiarasu L (2013) Use of plant extracts for the control of flacherie disease in silkworm, *Bombyx mori L.* (Lepidoptera: Bombicidae). Int J Microbiol Res 4(2):158–161

Laskowski RA, MacArthur MW, Moss DS, Thornton JM (1993) PROCHECK—a program to check the stereochemical quality of protein structures. J App Cryst 26:283–291

Maity S, Mukherjee K, Banerjee A, Mukherjee S, Dasgupta D, Gupta S (2016) Inhibition of porcine pancreatic amylase activity by sulfamethoxazole: structural and functional aspect. Protein J 35(3):237–246

Manimegalai S, Chandramohan N (2005) Botanicals for the management of bacterial flacherie of silkworm *Bombyx mori L.* Sericologia 45(1):55–58

Mukherjee K, Pandey DM, Vidyarthi AS (2014) In silico characterization and Analysis of RTBP1 and NgTRF1 protein through MD simulation and molecular docking—a comparative study. Interdiscip Sci Comput Life Sci. doi:10.1007/s12539-014-0237-6

Nataraju B, Sivaprasad V, Datta RK (1999) Studies on the cause of thatteroga in silkworms, *Bombyx mori L.* Indian J Seric 38:149–151

Patil BR, Singh KK, Pawar SE, Maarse L, Otte J (2009) Sericulture L: an alternative source of small-scale farmers and tribal communities, pre-poor livestock policy iniative—a living from livestock. Research Report RR Nr. 09-03

Rajitha K, Savithri G (2015) Studies on symptomological and economic parameters of silk cocoons of *Bombyx mori* inoculated with *BeauveriaBassiana* (Bals.) Vuill. Int J Curr Microbiol App Sci 4(2):44–54

Roy D, Mukherjee K (2015) Homology modeling and docking studies of human chitotriosidase with it's natural inhibitors. J Proteins Proteomics 6(2):183–196

Rungta D, Chauhan N, Mukherjee K (2014) Identification of Hiv1 protease inhibitor through molecular modelling and structure based virtual screening approach. J Adv Bioinf Appl Res 5 (3):140–149

Sakthivel S, Angaleswari C, Mahalingam PU (2012) Isolation and identification of bacteria responsible for flacherie in silkworms, Pelagia Research Library. Adv Appl Sci Res 3(6):4066–4068

Sastry GM, Day MAT, Sherman RAW (2013) Protein and ligand preparation: parameters, protocols, and influence on virtual screening enrichments, protein and ligand preparation: parameters, protocols, and influence on virtual screening enrichments. J Comput Aided Mol Des 27:221–234

Saxena VK, Sosanna A (2005) Siosterol-3-O-Beta-D xylopyranoside from the flowers of *Tridax procumbens* Linn. J Chem Sci 117(3):263–266

Selvakumar T, Nataraju B, Balavenkatasubbaiah M, Sivaprasad V, Baig M, Sharma kumar V, Thiagarajan SDV, Datta RK (2002) A report on prevalence of silkworm diseases and estimated crop loss. In: Dandin SB, Gupta VP (eds) Advances in Indian Sericulture Research. CSRTI, Mysore, pp 354–357

Van Der Spoel D, Lindahl E, Hess B, Groenhof G, Mark AE, Berendsen HJ (2005) GROMACS: fast, flexible, and free. J Comput Chem 26(16):1701–1718

Wood F (2002) The silk road: two thousand years in the heart of Asia. University of California Press, Berkeley, CA. pp 9, 13–23. ISBN 978-0-520-24340-8

Yang J, Roy A, Zhang Y (2013) Protein-ligand binding site recognition using complementary binding-specific substructure comparison and sequence profile alignment. Bioinformatics 29:2588–2595

Growth Phase-Dependent Synthesis of Gold Nanoparticles Using *Bacillus Licheniformis*

Swati Tikariha, Sharmistha Banerjee, Abhimanyu Dev and Sneha Singh

Abstract

Biological agents play an important role in biosynthesis of nanoparticles and regarded as green technology and environmental-friendly approach, which is an important step in the field of application of nanotechnology. The present study reports biosynthesis of gold nanoparticles from gold precursor using *Bacillus licheniformis* at 37 °C depending upon its growth phase. Growth phase study of bacteria was performed and biosynthesis of gold nanoparticles was carried out at early log phase, mid log phase, stationary phase, and decline phase. The synthesis of gold nanoparticles was confirmed by UV–visible spectrum that showed peak between 545 and 559 nm which was later reconfirmed by SEM studies. Size of gold nanoparticles as measured by dynamic light scattering was found to be in the range of 16–95 nm. Zeta potential measurement showed that gold nanoparticles were negatively charged with zeta potential value between −20.8 and −25.8 mV. The stationary phase culture, out of early and late log phase showed best results, synthesizing small-sized gold nanoparticles.

Keywords

Bacillus licheniformis · Biosynthesis · Gold nanoparticles · Growth phase · Zeta potential

S. Tikariha · S. Banerjee · S. Singh (✉)
Department of Bio-Engineering, BIT, Mesra, Ranchi, Jharkhand, India
e-mail: snehasingh@bitmesra.ac.in

A. Dev
Department of Pharmaceutical Sciences & Technology, BIT, Mesra, Ranchi, Jharkhand, India

K. Mukhopadhyay et al. (eds.), *Applications of Biotechnology for Sustainable Development*, DOI 10.1007/978-981-10-5538-6_15

Introduction

Recently, the synthesis of nanogold has become increasingly important owing to their unique properties that differ from their bulk counterparts (Armendariz et al. 2004). The nanosized gold particles find applications in diverse fields as catalysis, sensors and medicine which largely depend on the size and composition of the nanoparticles (Haverkamp et al. 2007). Synthesis of nanoparticles can be achieved through chemical, physical, and biological methods. As nanogold is widely used in vivo in human especially in contacting areas, there is a growing need to develop environmentally benign processes for their synthesis. As a result, biological approach of nanoparticles synthesis by using different microorganism, enzyme, and plant extract have emerged as possible alternatives to synthetic methods (Song and Kim 2009). A wide variety of microorganism has been used as a biofactory for biosynthesis of gold nanoparticles. Handling of bacteria is easy and they can be easily manipulated as per requirement, so bacterial system has gained importance for the synthesis of gold nanoparticles (Sweeney et al. 2006). The present study reports a simple biological synthesis of gold nanoparticles through *Bacillus licheniformis* by reduction of chloroaurate ions depending upon its different growth phase study. This work is the comprehensive kinetic study of gold nanoparticles aggregation that can be used to predict the appropriate growth phase of biomass culture required for biosynthesis of gold nanoparticles. Although, the kinetics of synthesis of gold nanoparticles following chemical and physical methods is available, there is very few information addressing the process of nucleation and final aggregation of gold nanoparticles by utilizing bacterial biomass from different growth phases. A deeper understanding of the underlying formation depending on different growth phase is still missing. So, the present study targets on growth phase-dependent biosynthesis of gold nanoparticles thus identifying the possible growth phase of *B. licheniformis* at 37 °C that will increase the understanding regarding rapid nucleation and growth processes of gold nanoparticles in solution phase. The mechanism by which different sized gold nanoparticles formed in different growth phase is still missing.

Materials and Methods

Growth Phase Study of *Bacillus Licheniformis*

Bacillus licheniformis MTCC 1502 was used for present work which was maintained on nutrient agar slants. To prepare biomass for growth phase study, *B. licheniformis* was cultured in Nitrate medium. The flasks were inoculated at 37 °C and agitated at 150 rpm. The cell growth in the broth was measured in triplicate at an equal interval of time and the standard plot of optical density of the cell suspension against time (h) was made to obtain the different growth phases of *B. licheniformis*.

The optical density of the cell suspension was noted using a spectrophotometer at 600 nm.

Growth Phase-Dependent Biosynthesis of Gold Nanoparticles and Its Characterization

For the growth phase-dependent biosynthesis of gold nanoparticles study, culture representing four different growth phases of *B. licheniformis* was taken. Culture was harvested after 5, 10, 24, and 48 h representing late lag or early log phase, mid log phase, stationary phase, and decline phase, respectively, by centrifugation at 5000 rpm for 20 min. The biomass thus obtained was washed two times with phosphate buffer of pH 7 to remove any media components that may interact with gold ions. The pellet so obtained was finally resuspended in distilled water followed by addition of auric chloride solution and kept in rotator shaker at 150 rpm, 37 °C for 24–48 h along with the control (only biomass) flask.

For the characterization of gold nanoparticles, the above reaction mixture was processed and characterized by UV/Visible spectroscopy (Perkin Elmer, Lambda-25, USA), Dynamic Light scattering (DLS, NanoZS, Malvern, UK), Zeta potential (NanoZS, Malvern, UK), and Scanning electron microscope (SEM, Jeol, JSM-6390LV, Japan) study.

Results and Discussion

Bacillus Licheniformis Growth Phase

A sigmoidal growth curve of *B. licheniformis* cultured in nitrate medium at 37 °C was obtained. The growth curve drawn between optical density of the cell suspension against time (hour) showed that cells started to grow after around 2–3 h of inoculation and up to 3 h, cells were in lag phase. Cells entered log or exponential phase after 3 h and continued up to 18 h followed by stationary phase that was continued upto 38 h after which, cells entered decline phase (Fig. 1).

Biosynthesis of Nanoparticles

Based on growth phase study, growth phase-dependent biosynthesis of gold nanoparticles was carried out. Gold nanoparticles were synthesized from auric chloride solution containing Au^{3+} ions by treating with the biomass pellet of early log phase (culture age of 5 h), mid log phase (culture age of 10 h), stationary phase (culture age of 24), and decline phase (culture age of 48 h). After 17 h of the reaction it was observed that the color of the solution in the flask containing early log phase, late log, and stationary phase biomass pellet turned to purple, red and

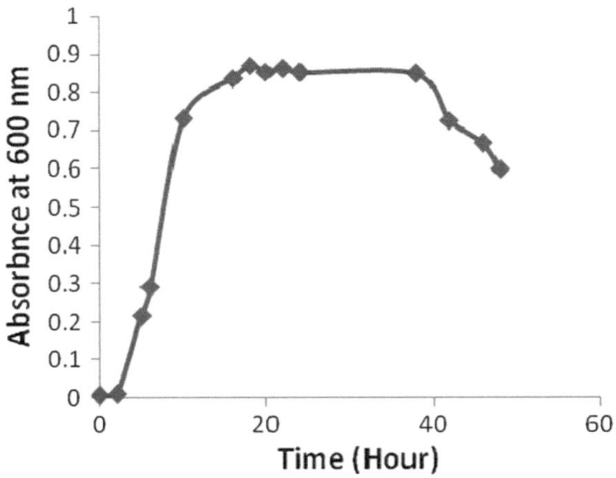

Fig. 1 Growth curve of *Bacillus licheniformis*

Fig. 2 Change in the color of reaction mixture to **a** *purple*, **b** *red* and **c** *purple* containing during early log phase (5 h), late log phase (10 h), stationary phase biomass (24 h) with respect to control flasks

purple, respectively, as shown in Fig. 2a–c. The nanoparticles thus synthesized were intracellular since the biomass were colored while the solution remained colorless. This confirms the formation of gold nanoparticles and the change in color was due to surface plasmon resonance. This important observation indicates the reduction of the Au^{3+} ions and the biosynthesis of gold nanoparticles. There was no change in the color of reaction mixture containing 48 h biomass pellet suggesting there might be degradation of the enzymes during the decline phase which are responsible for the bioreduction of Au(III) from solution to Au(0) as bacteria functions as a cellular efflux pumping system secreting proteins, enzymes, organic acids and polysaccharides and periplasmic protein, which binds gold specifically at the cell surface thereby causing alteration of solubility and toxicity via reduction, bio-sorption, bioaccumulation, and lack of specific metal transport systems (Beveridge and Murray 1980; Matias and Beveridge 2008).

UV–Visible Spectrum

UV–visible spectral analysis confirmed the biosynthesis of gold nanoparticles using *Bacillus licheniformis* of different growth phase. The UV–visible spectra showed a strong Plasmon resonance which was centered approximately at 545, 559, and 548 nm for early log phase, mid log phase, and stationary phase biomass respectively (Fig. 3a–c). This shows the formation of gold nanoparticles in solution and the peak position depends on several factors such as the particle size, shape, and dielectric constant of the medium. Observation of this strong broad plasmon peak has been well documented for various metal nanoparticles, with sizes ranging all the way from 2 to 100 nm (Sastry et al. 1998).

Dynamic Light Scattering

The size of the gold nanoparticles synthesized in different growth phases was estimated by dynamic light scattering study. The early log phase biomass synthesized gold nanoparticles in the size range of 29–82 nm. Maximum particles were of 76.36 nm as shown in Fig. 4a. Mid log biomass bioreduce auric chloride solution to gold nanoparticles in broad range between 34 and 95.07 nm. Maximum particles were of 65.92 nm (Fig. 4b). Gold nanoparticles of narrow size range were observed with stationary phase biomass. The nanoparticles were synthesized between 16 and 70.89 nm with maximum particles of 62.32 nm (Fig. 4c). The different sized nanoparticles so obtained are due to the level of expression of enzymes that are responsible for nucleation process followed by coalescence forming colloids. Stationary phase biomass, in terms of size range and maximum particle size, is synthesizing small-sized nanoparticles suggesting fast nucleation process leading to biosynthesis of small-sized nanoparticles due to enzyme production in sufficient amount. The mechanism of formation of gold nanoparticles is still not well elucidated, however, there are previous reports which claim that an enzyme belonging to the NADH reductase family is possibly involved (He et al. 2008; Kumar et al. 2008).

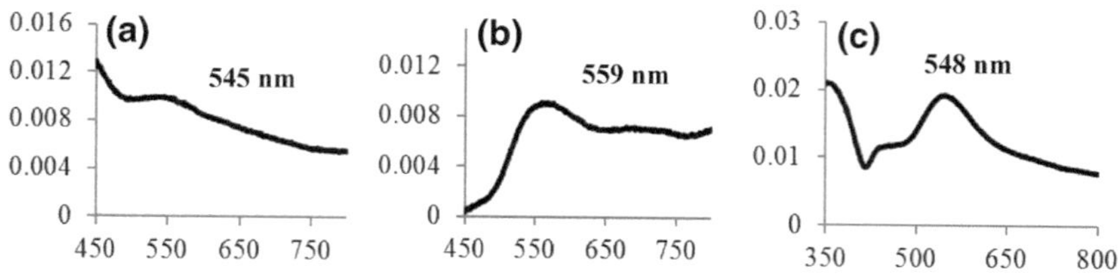

Fig. 3 Appearance of the localized surface plasmon bands in UV–Vis absorption spectra of reaction mixture containing **a** early log phase biomass, **b** mid log phase, and **c** stationary phase biomass respectively

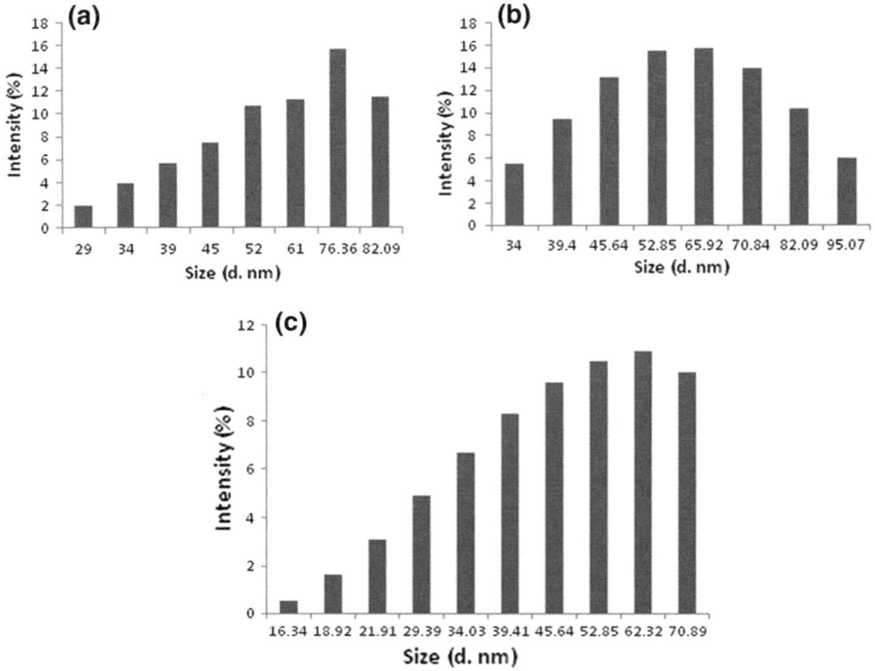

Fig. 4 Particles size distribution of gold nanoparticles synthesized using **a** early log phase, **b** late log phase, and **c** stationary phase biomass respectively

Zeta Potential

The high zeta potential value is an important parameter in maintaining the stability of any suspension because it creates a repulsive force which keeps the nanoparticles away from each other. Zeta potential measurement provides an indirect measure of the net charge. Zeta potential measurements at the end of the reduction reaction were negative. The net zeta potential of the gold nanoparticles synthesized using biomass pellet of *B. licheniformis* from early log, mid log and stationary phase was −25.8. −20.8 and −24.3 mV, respectively, suggesting nanoparticles synthesized from early log and stationary phase were more stable.

SEM Analysis

The SEM image was obtained as represented in Fig. 5a–c shows that most of the obtained nanoparticles were thought to have spherical shape in the range 60–146 nm.

EDS spectrum, linked with SEM, was used to analyze the element of gold nanoparticles (Fig. 6a–c). The EDS attachment on the SEM provided the chemical

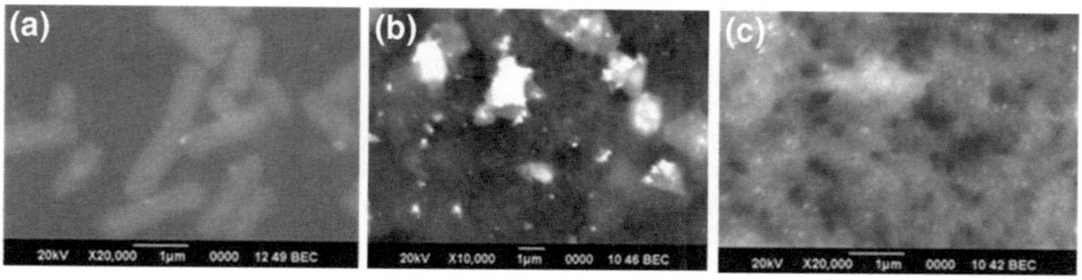

Fig. 5 SEM image of gold nanoparticles synthesized using **a** early log phase, **b** late log phase, and **c** stationary phase biomass respectively

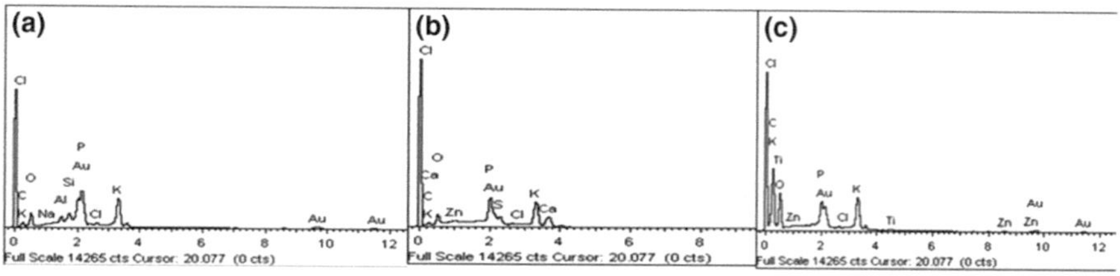

Fig. 6 EDS spectrum of gold nanoparticles synthesized using early **a** log phase, **b** late log phase, and **c** stationary phase biomass respectively

fingerprint of the field in view and the spot analysis of minute particles confirms the presence of specific elements. The SEM-EDS analysis displayed signature spectra for gold and thus convincingly evidenced the presence of this noble metal in the sample. These results are consistent with other reports on the EDS analysis of gold structures synthesized by using some fruit extracts.

Conclusion

The study demonstrates a simple procedure for the growth phase-dependent biosynthesis of gold nanoparticles through *B. licheniformis*, establishes a greener approach with several advantages such as cost effectiveness, compatibility for biomedical and pharmaceutical applications as well as for large-scale commercial production. It is anticipated that enzymes could have acted as the major bio-molecules involved in the synthesis and facilitated the synthesis of gold nanoparticles. In stationary phase, the enzymes responsible for bioreduction is thought to be secreted in more amounts than rest of growth phase as the size of nanoparticles was smaller ($\sim$62 nm) and in narrow range (16–70 nm) probably due to effective nucleation and growth of particles in aqueous solution. The zeta potential also suggests the stability of nanoparticles in aqueous solution. With this, we can conclude that stationary phase culture of *B. licheniformis* at 37 °C can be directly used for effective gold nanoparticles biosynthesis.

References

Armendariz V, Herrera I, Jose-yacaman M, Troiani H, Santiago P, Gardea-Torresdey JL (2004) Size controlled gold nanoparticle formation by *Avena sativa* biomass: use of plants in nanobiotechnology. J Nanopart Res 6(4):377–382

Beveridge TJ, Murray RG (1980) Sites of metal deposition in the cell wall of *Bacillus subtilis*. J Bacteriol 141(2):876–887

Haverkamp RG, Marshall AT, van Agterveld D (2007) Pick your carats: nanoparticles of gold–silver–copper alloy produced in vivo. J Nanopart Res 9(4):697–700

He S, Zhang Y, Guo Z, Gu N (2008) Biological synthesis of gold nanowires using extract of *Rhodopseudomonas capsulata*. Biotechnol Prog 24(2):476–480

Kumar SA, Peter YA, Nadeau JL (2008) Facile biosynthesis, separation and conjugation of gold nanoparticles to doxorubicin. Nanotechnology 19(49):495101

Matias VR, Beveridge TJ (2008) Lipoteichoic acid is a major component of the *Bacillus subtilis* periplasm. J Bacteriol 190(22):7414–7418

Sastry M, Patil V, Sainkar SR (1998) Electrostatically controlled diffusion of carboxylic acid derivatized silver colloidal particles in thermally evaporated fatty amine films. J Phys Chem B 102(8):1404–1410

Song JY, Kim BS (2009) Rapid biological synthesis of silver nanoparticles using plant leaf extracts. Bioprocess Biosyst Eng 32(1):79–84

Sweeney SF, Woehrle GH, Hutchison JE (2006) Rapid purification and size separation of gold nanoparticles via diafiltration. J Am Chem Soc 128(10):3190–3197

A Rapid Method for Detection and Characterization of Anthocyanins from *Hibiscus*, *Ocimum* and *Syzygium* Species and Evaluation of Their Antioxidant Potential

Biswatrish Sarkar, Manish Kumar and Kunal Mukhopadhyay

Abstract

Ocimum tenuiflorum, *Hibiscus rosa-sinensis*, *Hibiscus sabdariffa* and *Syzygium cumini* have been known traditionally for their health-promoting properties. These beneficial health-promoting effects are strongly attributed to the flavonoids and anthocyanins content of these plants. The objective of the study was to develop an efficient analytical method for rapid identification of anthocyanins and evaluate their antioxidant properties. Characterization of the anthocyanins was done by directly infusing the anthocyanin extracts into an ESI-IT/MS/MS in positive ion mode. Antioxidant activity of the anthocyanins were evaluated by 2,2-diphenyl-1-picrylhdrazyl and hydroxyl radical scavenging assay. ESI-IT/MS/MS study helped to characterize the anthocyanins by revealing their location of glycosidation and acylated groups. The extracted anthocyanins showed excellent antioxidant activity in vitro. The in vitro antioxidant studies revealed the capacity of anthocyanins as free radical scavengers. The analytical approach developed in this study can be further applied for the profiling of metabolomics and investigating effect of other factors such as cultivars, time of harvesting and seasonal variation on anthocyanin biosynthesis in *Hibiscus*, *Ocimum* and *Syzigium*.

Keywords

ESI-IT/MS/MS · Anthocyanins · Antioxidant assay · Hydroxyl radical scavenging · *Hibiscus* · *Ocimum* · *Syzigium*

B. Sarkar
Department of Pharmaceutical Sciences & Technology,
Birla Institute of Technology, Mesra, Ranchi 835215, Jharkhand, India

M. Kumar · K. Mukhopadhyay (✉)
Department of Bio-Engineering, Birla Institute of Technology,
Mesra, Ranchi 835215, Jharkhand, India
e-mail: kmukhopadhyay@bitmesra.ac.in

K. Mukhopadhyay et al. (eds.), *Applications of Biotechnology for Sustainable Development*, DOI 10.1007/978-981-10-5538-6_16

Introduction

Anthocyanins, like flavonoids, continue to capture the interest of biologists because of their structural diversity, ecological significance and health-promoting effects. It imparts purple, red and blue colour to leaves, fruits and flowers in different plants. Chemically, anthocyanins are *O*-glycosides of polyhydroxy or polymethoxy derivatives of 2-Phenylbenzopyrilium; the aglycones are known as anthocyanidins (Kong et al. 2003; Bueno et al. 2012). Consumption of anthocyanins contributes towards prevention of cardiovascular diseases, cancer and diabetes (Kong et al. 2003; Konczak and Zhang 2004; Wallace 2011; Bueno et al. 2012). These health-promoting effects are strongly attributed to its antioxidant property (Shipp and Abdel-Aal 2010; Santos-Buelga et al. 2014). With the increasing concern about food safety and human health, natural and plant-derived pigments like anthocyanins deserve more attention (Vyas et al. 2009). Same time it is essential to develop newer techniques for rapid extraction, separation and structural analysis for progress in research on anthocyanins (Wu and Prior 2005; Valls et al. 2009).

Ocimum tenuiflorum L. (Lamiaceae), *Hibiscus rosa-sinensis* L. (Malvaceae), *Hibiscus sabdariffa* L. (Malvaceae) and *Syzygium cumini* L. (Myrtaceae) were reported to be repositories of various therapeutic phytochemicals like flavonoids and anthocyanins (Sarkar et al. 2014). The phyto-constituents derived from these plants have demonstrated potential pharmacological effects and health benefits like expectorant, analgesic, cardiovascular, antiemetic, diaphoretic, antidiabetic, antifertility, hepatoprotective, hypotensive, hypolipidemic, antistress, antiapoptotic and anticancer properties (Khanna and Bhatia 2003; Sarkar et al. 2014). Many of these properties have been directly attributed to the anthocyanin content of these plants. This makes these plants an ideal choice for anthocyanin research.

Structural diversity and substitution pattern is a challenge for direct analysis of anthocyanins. Notably improvement has been observed in direct analysis of anthocyanidins using variety of hyphenated techniques like LC-MS, LCNMR, etc. (Gao and Mazza 1994; Durst and Wrolstad 2001; Nyman and Kumpulainen 2001; Burdulis et al. 2008). Among the different tandem mass spectrometers, electrospray ionization with ion trap mass spectrometers (ESI-IT/MS) and quadrupole time of flight mass spectrometers (Q-TOF) are extensively used for characterization of anthocyanins. ESI-ITMS uses sequential fragmentation (MS^n) presenting low mass resolution with high sensitivity and high mass accuracy for both precursor and product ions.

In the present study, characterization of anthocyanins from the selected plants was carried out by directly infusing these plant extracts to an advanced analytical platform of ESI-IT/MS/MS. The free radical scavenging capacity of the anthocyanin extracts were also estimated to determine their antioxidant potential.

Materials and Methods

Plant Material

Leaves of *O. tenuiflorum* L. (Lamiaceae), petals of *H. rosa-sinensis* L. (Malvaceae), calyx of *H. sabdariffa* L. (Malvaceae) and fruits of *S. cumini* L. (Myrtaceae) were code named OT, HRS, HS and SC, respectively, and collected from plants grown in the Indigenous Medicinal Plants Garden of Birla Institute of Technology, Mesra, Ranchi, India (23° 24′N, 85° 26′E, 619 m asl). The plant materials were collected during the month of November and dried under shade.

Chemicals and Reagents

All chemicals were of analytical reagent grade. Acetonitrile (LC-MS hypergrade, 99.9%) and LC grade Methanol, Ortho-phosphoric acid, Hydrochloric acid were purchased from Merck (Darmstadt, Germany). Water used in all procedures was purified through a MilliQ system (Millipore, Bedford, MA, USA). 2,2-diphenyl-1-picrylhydrazyl (DPPH), trichloroacetic acid (TCA), thiobarbituric acid (TBA), 2-deoxy-D-ribose and quercetin were purchased from Sigma-Aldrich, Germany.

Characterization of Anthocyanins by Direct Infusion to ESI-IT/MS/MS

Extraction of Anthocyanins

Extraction of the anthocyanins from powdered plant parts as mentioned earlier were performed according to of Sarkar et al. (2014). The anthocyanin extracts (AEs) were collected, stored at 4 °C and used for future studies.

Analytical Conditions

For characterization of anthocyanins the AEs (10 µg) were dissolved in 1 mL solvent containing methanol: 0.1% formic acid in water. All experiments were conducted on a high capacity ion trap ultra (HCT Ultra, PTM discovery) in ESI positive ion mode using an Esquire 3000-plus mass spectrometer (Bruker Daltonics, Bremen, Germany). The instrument was used in full scan mode to detect and acquire an overview of all abundant metabolites present in the sample. Ions were detected in ion charge control (ICC) mode at *m/z* ranging 50–500 µm. Positive ions were detected at unit resolution (scan speed 13,000 µm/s), 10 scans were summarized for one spectrum, resulting in a spectral rate of 1 Hz. The following conditions were maintained during the experiment: ESI interface, nebulizer gas pressure, 10 psi; target mass 500 m/z; Capillary voltage, 3 kV in positive mode;

drying gas flow rate, 4 L/min; drying gas temperature, 300 °C; MS/MS, scan from m/z 15 to 3000; ion trap, scan from m/z 50 to 1800; maximum accrual time, 200,000 μs; average, 10; smart parameter setting (SPS), compound stability, 100%; trap drive level, 100%. The data were processed using Esquire data analysis software, version 3.1 (Bruker Daltonics, Bremen, Germany).

Antioxidant Activity of Anthocyanins

DPPH (2,2-Diphenyl-1-Picrylhydrazyl) Assay

The scavenging of free radical activity of the AEs was quantified by the quenching of DPPH free radicals (Sarkar et al. 2014). Quercetin was used as positive control and standard in this experiment. Percent inhibition (*I*) was calculated by the following equation and IC_{50} was calculated from the regression analysis.

$$I\ (\%) = \left[\left(A_{\mathrm{control}} - A_{\mathrm{sample}}\right) / A_{\mathrm{control}}\right] \times 100\%$$

Hydroxyl Radical Scavenging Assay

The scavenging of the hydroxyl radical ($OH^{\bullet}$) by AEs were measured by modification of the standard deoxyribose method (Haliwell et al. 1987). The following two experimental variations were performed to establish (i) role of antioxidants on hydroxyl trapping (non-site-specific assay) and (ii) role of antioxidants on metal chelation (site-specific assay).

Statistical Analysis

For antioxidant assays all data were reported as mean ± standard deviation ($n = 3$) and evaluated by one-way analysis of variance (ANOVA) followed by Tukey's multiple comparison tests. Differences were considered statistically significant at the value of probability less than 5% ($p < 0.05$).

Results and Discussion

Characterization of Anthocyanins by Direct Infusion to ESI-IT/MS/MS

ESI/MS/MS technique is a useful and powerful technique for characterization of anthocyanins. This ionization method is very sensitive and produces intact molecular ions with corresponding fragments of anthocyanidins. The fragmentation patterns assist in determining the location of glycosides and acylation in an anthocyanin compound (Giusti et al. 1999). The ESI/MS/MS technique and direct infusion of anthocyanin extracts to the mass spectrometer provide certain advantages like ease of sample preparation, short run time and a consistent anthocyanin profile. The direct infusion method requires no prior separation of the extracts by column chromatography or gradation of mobile phases and same time provides more reproducible results (Giusti et al. 1999).

Accordingly, in this study, the AEs were characterized by direct infusion into ESI-IT/MS/MS which resulted in clear and characteristic fragmentation pattern of anthocyanins. The positive ESI mode effected in dissociation of glycosidic bonds (at positions of C3 and C5 of anthocyanidins) but only between the flavylium skeleton and sugars attached to it. For example, the cleavage produced at 3 glycosylated anthocyanins (Cyanidin-3-glucoside, *m/z* 448.9) produced fragment that corresponds to the aglycon, the C3 substituted anthocyanin (i.e. Cyanidin, *m/z* 287.1) (Fig. 1a). Similarly, MS of 3,5-glycosylated anthocyanin (i.e. Cyanidin-3,5-diglucoside, *m/z* 610.8) produced fragments corresponding to C3 substituted anthocyanin (Cyanidin, *m/z* 287.1) and C5 substituted anthocyanin (Cyanidin-3-glucoside, *m/z* 448.9), along with molecular parent ion (Fig. 1b). Thus, visibly the detection was based on *m/z* of molecular ion $[M]^+$ and fragment ions along with the dissociation of glycoside attached to an aglycon of AEs. The AEs, as evident from ESI-IT-MS/MS, contained wide variety of anthocyanins with different acylation pattern and glycons (Table 1). The affirmation of the structures of individual anthocyanins was made based on data of ESI/MS/MS from previously published literatures (Giusti et al. 1999). In short, the direct infusion of AEs into ESI/MS/MS provided a comprehensive data of anthocyanins present in all the plant species and their substitution patterns.

Antioxidant Activity of Anthocyanins

For assessing the free radical scavenging activity, bleaching of DPPH was carried out with increasing concentrations of AEs and was compared with quercetin as standard. All the AEs and quercetin scavenged DPPH free radicals in a concentration-dependent fashion (Fig. 2a). One-way ANOVA followed by Tukey's multiple comparison tests among the IC_{50} values indicated significant difference ($p < 0.05$) in activities between different AEs. SC has the best antioxidant activity among the four species as it demonstrated low IC_{50} value (24.738 ± 2.23 μg/mL) and was equivalent to standard quercetin (20.413 ± 1.97 μg/mL). HS showed highest IC_{50} value among the four species that proved its low scavenging property ($IC_{50} = 63.878 \pm 3.09$ μg/mL) (Fig. 3a).

The antioxidant activity of AEs were also evaluated by reduction of hydroxyl radical ($OH^{\bullet}$) mediated 2-deoxy-D-ribose chromogen. Alike the previous assay, all AEs showed concentration-dependent hydroxyl radical scavenging activity (Fig. 2b, c). SC showed best results in both site-specific and non-site-specific 2-deoxy-D-ribose degradation assay (Fig. 3b, c) and its IC_{50} value (36.777 ± 0.78 μg/mL) was

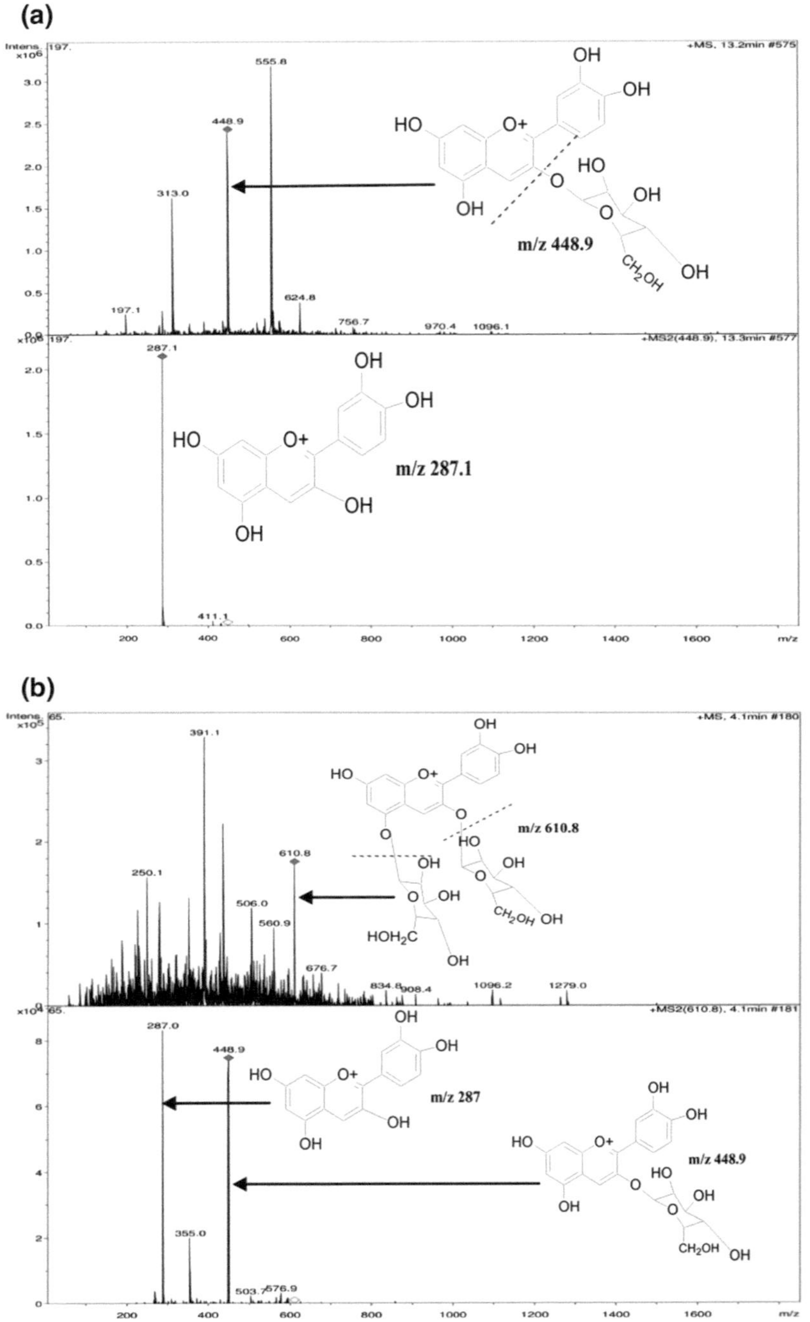

Fig. 1 ESI-MS/MS fragmentation pattern and proposed structures of few of the compounds. **a** Cyanidin-3-glucoside. **b** Cyanidin-3,5-diglucoside

Table 1 Identification of anthocyanins by ESI-IT-MS/MS

Compound name	Molecular ion-M$^+$ (*m/z*)	Fragment ion-MS/MS (*m/z*)
Hibiscus sabdariffa		
Delphinidin 3-(6″-coumaroyl) glucoside	610.8	303
Delphinidin 3-sambubioside	596.8	303
Delphinidin 3-galactoside	464.9	303
Cyanidin-3-sambubioside	580.8	287
Cyanidin 3-(6″-coumaroyl) glucoside	594.8	287
Peonidin-3-acetylglucoside	504.9	301
Hibiscus rosa-sinensis		
Cyanidin-3,5-diglucoside	610.8	287.1, 449
Peonidin-3-(6″-acetoyl) glucoside	504.9	301
Ocimum tenuiflorum		
Cyanidin-3,5-diglucoside	610.8	287, 448.9
Cyanidin-3-glucoside	448.9	287.1
Cyanidin3-(6″-acetoyl) galactoside	490.9	287.1
Cyanidin 3-(6″-coumaroyl) glucoside	594.9	287.1
Peonidin-3-(6″-acetoyl) galactoside	504.9	301
Quercetin 3-*O*-glucuronide	476.9	301
Syzygium cumini		
Cyanidin-3,5-diglucoside	610.8	287, 448.9
Cyanidin-3,6-acetylgalactoside	490.9	287
Delphinidin 3-galactoside	465	303.1
Malvidin-3,5-diglucoside	654.8	331, 492.9
Petunidin 3,5-diglucoside	640.8	317.1, 478.9

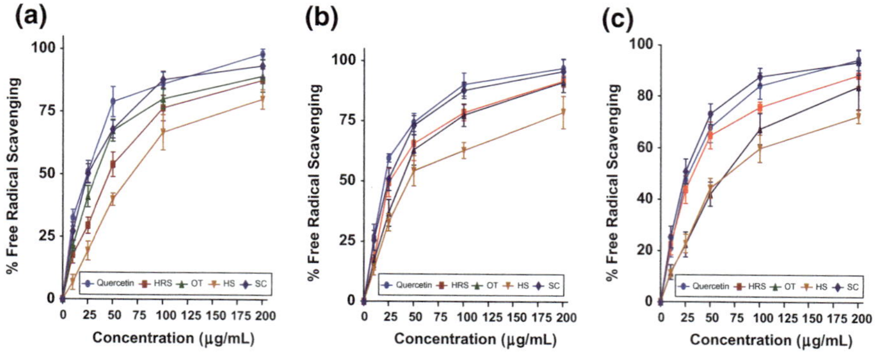

Fig. 2 Concentration-dependent free radical scavenging activity of anthocyanin extracts. **a** DPPH assay, **b** non-site-specific OH radical scavenging assay, **c** site-specific OH radical scavenging assay (Data presented as mean ± SD, $n = 3$; HS: *Hibiscus sabdariffa*, HRS: *Hibiscus rosa-sinensis*, OT: *Ocimum tenuiflorum*; SC: *Syzygium cumini*)

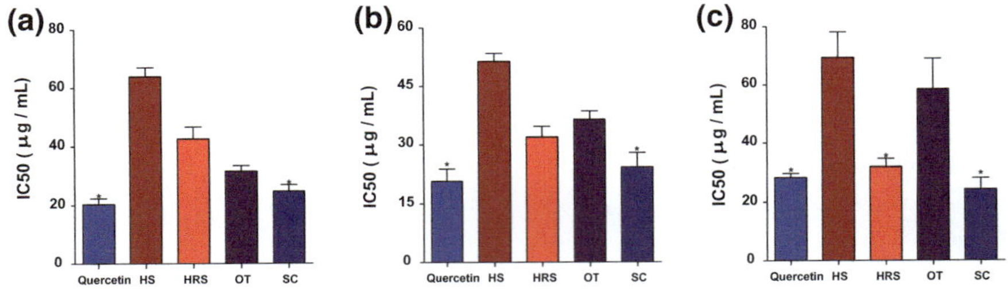

Fig. 3 Quantitative comparison of IC_{50} values of in vitro antioxidant assays of anthocyanin. **a** DPPH free radical scavenging assay, **b** non-site-specific OH radical scavenging assay, **c** site-specific OH radical scavenging assay (Data presented as mean ± SD, for $n = 3$); $p < 0.05$ compared among the test groups as well as with standard (quercetin). *$p < 0.05$ indicate no significant difference between the experimental group and standard values (HS: *Hibiscus sabdariffa*, HRS: *Hibiscus rosa-sinensis*, OT: *Ocimum tenuiflorum*; SC: *Syzygium cumini*)

statistically analogous with quercetin (calculated Tukey Q value = 0.1721 whereas critical Tukey Q value = 4.53, $n = 4$, error $df = 11$).

Antioxidant activity of anthocyanins has been widely documented. The *O*-dihydroxy structure in its B ring, as in cyanidin, confers stability to the radical form and participates in electron delocalization. The hydroxyl groups in anthocyanins donate hydrogen to free radicals, stabilize them and give rise to relatively stable flavonoids phenoxyl radicals. The electron transfer also neutralizes the free radicals. Anthocyanins have been reported to contribute greater towards free radical scavenging than other phenolics (Bi et al. 2014). AEs from *Hibiscus* and *Ocimum* have shown excellent antioxidant activity in our earlier study (Sarkar et al. 2014). Whereas, *Syzigium* AEs proved to be even better antioxidants in this study owing to presence of all the six major anthocyanins in it.

Summary

The direct infusion of extracts and ease of ESI/MS/MS application yielded excellent information regarding the exact mass, fragmentation patterns and molecular structures of anthocyanins. The in vitro antioxidant studies once again revealed the versatility of anthocyanins as a free radical scavenger. The analytical approach developed in this study can be further applied for the profiling of metabolomics and investigating effect of other factors such as cultivars, time of harvesting and seasonal variation on anthocyanin biosynthesis in *Hibiscus*, *Ocimum* and *Syzigium* species.

References

Bi X, Zhang J, Chen C, Zhang D, Li P, Ma F (2014) Anthocyanin contributes more to hydrogen peroxide scavenging than other phenolics in apple peel. Food Chem 152:205–209

Bueno JM, Sáez-Plaza P, Ramos-Escudero F, Jiménez AM, Fett R, Asuero AG (2012) Analysis and antioxidant capacity of anthocyanin pigments. Part II: chemical structure, color, and intake of Anthocyanins. Crit Rev Anal Chem 42:126–151

Burdulis D, Janulis V, Milašius A, Jakštas V, Ivanauskas. (2008) Method development for determination of anthocyanidin content in bilberry (*Vaccinium myrtillus* L) fruits. J Liq Chromatogr Relat Technol 31:850–864

Durst RW, Wrolstad RE 2001 Separation and characterization of anthocyanins by HPLC. Curr Protoc Food Analyt Chem F1.3.1–F1.3.13

Gao L, Mazza G (1994) Quantitation and distribution of simple and acylated anthocyanins and other phenolics in blueberries. J Food Sci 59:1057–1059

Giusti MM, Rodrigues-Saona LE, Griffin D, Wrolstad RE (1999) Electrospray and tandem mass spectroscopy as tools for anthocyanin characterization. J Agric Food Chem 47:4657–4664

Halliwell B, Gutteridge JMC, Aruoma OI (1987) The deoxyribose method: a simple "test-tube" assay for determination of rate constants for reactions of hydroxyl radicals. Anal Biochem 165:215–219

Khanna N, Bhatia J (2003) Antinociceptive action of *Ocimum sanctum* (Tulsi) in mice: possible mechanisms involved. J Ethnopharmacol 88:293–296

Kong JM, Chia LS, Goh NK, Chia TF, Brouillard R (2003) Analysis and biological activities of anthocyanins. Phytochemistry 64:923–933

Konczak I, Zhang W (2004) Anthocyanins- more than nature's colours. J Biomed Biotechnol 5:239–240

Nyman NA, Kumpulainen JT (2001) Determination of anthocyanidins in berries and red wine by high-performance liquid chromatography. J Agric Food Chem 49:4183–4187

Santos-Buelga C, Mateus N, De Freitas V (2014) Anthocyanins. plant pigments and beyond. J Agric Food Chem 62:6879–6884

Sarkar B, Kumar D, Sasmal D, Mukhopadhyay K (2014) Antioxidant and DNA damage protective properties of anthocyanin-rich extracts from *Hibiscus* and *Ocimum*: a comparative study. Nat Prod Res 28:1393–1398

Shipp J, Abdel-Aal EM (2010) Food applications and physiological effects of anthocyanins as functional food ingredients. The Open Food Sci J 4:7–22

Valls J, Millán S, Martí MP, Borrás E, Arola L (2009) Advanced separation methods of food anthocyanins, isoflavones and flavanols. J Chromatogr A 1216:7143–7172

Vyas P, Chaudhary B, Mukhopadhyay K, Bandopahyay R (2009) Anthocyanins: looking beyond colors. In: Bhowmik PK, Basu SK, Goyal A (eds) Advances in biotechnology, 1st edn. Bentham Science Publishers Ltd., Oak Park, pp 152–184

Wallace TC (2011) Anthocyanins in cardiovascular disease. Adv Nutr 2:1–7

Wu X, Prior RL (2005) Systematic identification and characterization of anthocyanins by HPLC-ESI-MS/MS in common foods in United States: fruits and berries. J Agri Food Chem 53:2589–2599

Phytochemical Screening and Antioxidant Property of Anthocyanins Extracts from *Hibiscus rosa-sinensis*

Akancha Anand and Biswatrish Sarkar

Abstract

This study was aimed at investigating the antioxidant property of anthocyanin extracts (AEs) from *Hibiscus rosa-sinensis* (petal). Preliminary phytochemical studies confirmed the presence of tannins, saponins, alkaloids, steroids, flavonoids and carbohydrates in the extract. Total phenolic content (TPC), total flavonoid content (TFC) and total anthocyanin content (TAC) of the AEs were also determined. In vitro antioxidant activity of the AEs was assessed by nitric oxide (NO) scavenging assay. The AEs demonstrated efficient and dose dependent inhibition of nitric oxide. A strong correlation was observed between antioxidant activity when compared with TPC, TFC and TAC. This confirmed that phenolics, flavonoids and anthocyanins were responsible for the exhibited antioxidant activity. Thus, this study highlighted *H. rosa-sinensis* having antioxidant properties which may be due to its anthocyanins, phenolics and flavonoids contents.

Keywords

Hibiscus rosa-sinensis · Antioxidant potential · Total phenol content · Total flavonoid content · Total anthocyanin content · Scavenging activity

A. Anand · B. Sarkar (✉)
Department of Pharmaceutical Sciences & Technology, Birla Institute of Technology, Mesra, Ranchi 835215, Jharkhand, India
e-mail: biswatrishsarkar@bitmesra.ac.in

A. Anand
e-mail: akanshaanand10@gmail.com

K. Mukhopadhyay et al. (eds.), *Applications of Biotechnology for Sustainable Development*, DOI 10.1007/978-981-10-5538-6_17

Introduction

In recent times the term 'free radical' such as reactive oxygen species (ROS) and reactive nitrogen species (RNS) has gained notoriety. Free radicals initiate chain reactions affecting biomolecules like lipids, proteins, etc. (Harman et al. 1956; Halliwell et al. 1997). ROS and RNS are the potential sources of pathogenesis of several diseases from Alzheimer's to cancer. 'Antioxidants' are the substances that neutralize free radicals (Halliwell et al. 1997) and may be obtained from number of sources like dietary supplements derived from plants. Anthocyanin is a plant-derived flavonoid which has gained acclaim as potent antioxidant in recent times. Studies suggested that the anthocyanins have a protective role against various diseases owing to their antioxidant activities (Soto-Vaca et al. 2012).

The dark red flowers of *H. rosa-sinensis* (Malvaceae) are important source of anthocyanins such as cyanidin-3-glucoside (Chia et al. 2003). In Ayurveda and other ancient literatures, *H. rosa-sinensis* has been used in various conditions like hypertension, pyrexia, liver disorders, antifertility activity, epilepsy, leprosy, bronchial catarrh and diabetes (Kumar et al. 2012).

The antioxidant potential of anthocyanin extracts (AEs) of *H. rosa-sinensis* was evaluated and compared with natural antioxidant in this study. The phytochemical components of AEs were identified by preliminary phytochemical screening. For this purpose, total phenolics contents (TPC), total monomeric anthocyanin contents (TAC) and total flavonoid contents (TFC) were also estimated. To assess the antioxidant property of AEs, nitrous oxide scavenging activity was performed and compared to that of ascorbic acid. The total polyphenol content, anthocyanin content and flavonoid content were correlated with antioxidant capacity as well, to evaluate their contribution towards total antioxidant function.

Materials and Methods

Chemicals

Folin–Ciocalteu reagent, anhydrous aluminium chloride, ferric chloride, ninhydrin reagent, sodium nitrite, gallic acid, quercetin, sodium carbonate, sodium nitroprusside, ascorbic acid sulfanilamide, naphthylethylenediamine dihydrochloride, potassium ferricyanide and trichloroacetic acid, were obtained from Sigma-Aldrich (Germany). All other used chemicals and solvents were of the high analytical grade. All aqueous solutions were prepared using water purified through MilliQ system (Millipore, Bedford, MA, USA).

Collection of Plant Materials

Flower of *H. rosa-sinensis* L. (Malvaceae) was code-named *HRS* and was collected from plants grown in the indigenous Medicinal Plant's Garden of Birla Institute of Technology, Mesra, Ranchi, India during the month of August and September. The collection of flower was based on colour of the flower and number of petals. Matured red flowers of five petals from a single plant were collected, calyx and anther parts were separated. The separated petals were dried under shade and powdered.

Preparation of Anthocyanin Extracts

Extraction of the pigments was carried out as previously described (Wang et al. 2000). The AEs were precipitated and the crude dried extract was stored at 4 °C for further investigations. The collected AEs were termed as HRSAEs in this study.

Phytochemical Screening

A small portion of the dry extract was used for the phytochemical tests for compounds which include tannins, flavonoids, alkaloids, saponins, steroids, carbohydrates, in accordance with the standard methods (gelatin test, Shinoda test, Alkaline reagent test, lead acetate solution test, ammonium test, Wagner's test, Dragendroff's test, Hager's test, Mayer's test, Foam test, Molisch's test, Benedict's test, respectively) with little modifications (Tiwari et al. 2011; Prasad et al. 2014).

Estimation of Total Phenolic Content (TPC)

The total phenolic content of the flower extracts were determined based on the folin–ciocalteu (FC) method (Singleton et al. 1965). All the results were expressed as mg gallic acid equivalents (GAE) per gram of sample.

Estimation of Total Flavonoids Content (TFC)

Total flavonoids in the sample extracts were determined using the aluminium chloride method (Liu et al. 2008). Results of the total flavonoid content were expressed as mg quercetin equivalents (QE) per 100 g of dry weight of the sample.

Estimation of Total Monomeric Anthocyanin Content (TAC)

Total monomeric anthocyanin was determined by pH differential method (Giusti et al. 2001). The TAC was reported as mg of anthocyanins/100 g dry weight (mg cyanidin-3-glucoside/100 g dry weight).

In Vitro Nitric Oxide Scavenging Assay

Nitric oxide radical inhibition was estimated using Griess Illosvory reaction with little modification (Patel et al. 2010). Ascorbic acid used as standard. The percentage inhibition was calculated using the formula:

$$\% \text{ scavenging activity } = [(A_{\text{control}} - \text{or } A_{\text{test}} \text{ or } A_{\text{std}})/A_{\text{control}}] \times 100$$

Statistical Analysis

All data were reported as mean ± standard deviation (for $n = 3$). Pearson correlation coefficients (r) and regression coefficient (R^2) were used to show correlations (at 95% confidence interval, where $p < 0.0001$).

Results and Discussion

The results of phytochemical screening of AEs indicated presence of tannins, saponins, alkaloids, steroids, flavonoids and carbohydrates (Table 1). Our result was in agreement with previous reports in same species showing presence of these phytochemicals (Bhaskar et al. 2011; Faten et al. 2012).

The total phenolic content of HRSAE was found to be 40.25 mg/g GAE (Table 2). The presence of polyphenolic compounds in the plant contributes to its colour, flavour and protective properties (Bueno et al. 2012).

A standard plot of quercetin ($y = 0.012x$, $R^2 = 0.954$) was plotted, Fig. 1b, and the total flavonoid content of HRSAE was found to be 35.33 mg QE/100 gm of dry weight (Table 2). Flavonoids are the most important natural phenols which are used by plant against oxidation (Andersen and Markham 2006). The estimation of TFC could provide an insight towards antioxidant properties of *HRS* flowers.

The TAC of the HRSAEs was found out to be 158 mg of anthocyanins/100 g dry weight (mg cyanidin-3-glucoside 100/g dry weight) (Table 2). Anthocyanins are excellent antioxidants known to possess various pharmacological properties (Kong et al. 2003).

Table 1 The phytochemical screening

S. No.	Phytochemical constituents	Tests	Presence
1.	Tannins	Gelatin test	++
2.	Phenols	Ferric chloride test	+++
3.	Alkaloids	(a) Mayer's test (b) Wagner's test (c) Hager's test (d) Dragendroff's test	++ ++ ++ ++
4.	Flavonoids	(a) Alkaline reagent test (b) Lead acetat solution test (c) Shinoda test (d) Ammonium test	+++ +++ +++ +++
5.	Saponins	Foam test	+
6.	Steroids		++
7.	Carbohydrates	(a) Molisch's test (b) Benedict's test	– +

+++ = appreciable amount (positive within 5 min); ++ = moderate amount (positive after 5 min but within 10 min); + = trace amount (positive after 10 min but within 15 min)

Table 2 Estimation of TPC, TFC and TAC of AEs from *Hibiscus rosa-sinensis*

Total phenolic content (mg/gm GAE)	Total flavonoid content (mg QE/100 gm of dry weight)	Total monomeric anthocyanin content (mg/100 g dry weight)
40.25	35.33	158

Nitric oxide degrades under the aerobic condition and forms nitrates and nitrites by reacting with oxygen, which could be estimated using the Griess reagent. The study is based on detection of decrease in nitrous acid in presence of a radical scavenger (Patel et al. 2010). Here HRSAEs showed strong concentration-dependent inhibitory/scavenging effect of nitric oxide radical (Fig. 2 and Table 3). The IC_{50} value calculated by regression analysis was found to be 62.25 μg/ml for HRSAE as compared to the standard ascorbic acid value, i.e., 35.28 g/ml (Table 4). A maximum inhibitory effect was seen at 40 μg/ml of the extract. The toxicity and damage caused by NO and O_2 are multiplied as they react to produce reactive peroxynitrile (ONOO–), which leads to serious toxic reactions with biomolecules. Nitric oxide is implicated in inflammation, cancer and other pathological conditions (Moncada et al. 1991; Lee et al. 2007). Our result confirms the excellent potential of HRSAE to counteract free radicals like NO.

To compare the antioxidant potential of the HRSAE, results of antioxidant capacities were correlated with TPC, TFC and TAC. Significant and high correlation between NO scavenging and TPC, YFC and TAC was observed (Fig. 3 and Table 5). Similar correlation between TPC, TFC and antioxidant activity has been reported in studies with other plants (Thaipong et al. 2006). These correlation comparisons exhibit linearity between antioxidant potential and TPC, TFC and

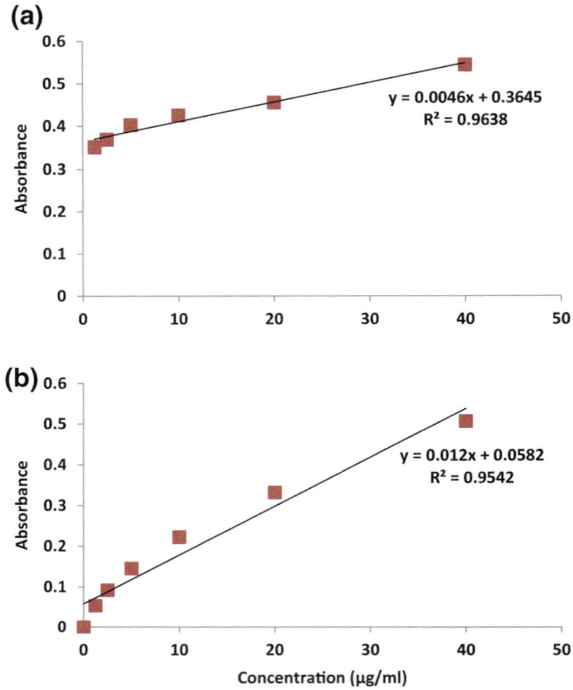

Fig. 1 Calibration curves for estimation of **a** total phenolic content (mg/g GAE) and total flavonoid content (mg QE/100 g of dry weight) of AEs

Table 3 Concentration dependent NO scavenging activity of AEs from *Hibiscus rosa-sinensis*

Concentration (μg/mL)	Ascorbic acid[a]	HRSAE[a]
1.25	5.670 ± 1.87	4.928 ± 0.29
2.5	9.757 ± 1.09	6.777 ± 1.45
5	17.931 ± 6.40	12.423 ± 1.59
10	30.744 ± 7.97	22.587 ± 4.35
20	42.563 ± 4.68	29.568 ± 0.39
40	48.196 ± 2.34	38.090 ± 2.75

[a]*Note* Values are as mean ± SD, $n = 3$

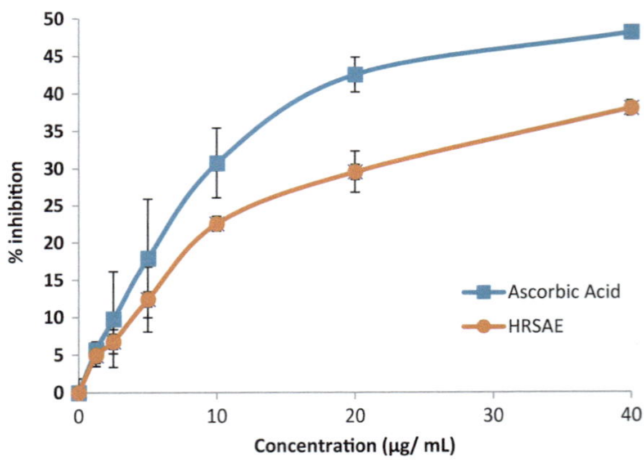

Fig. 2 Concentration dependent free radical scavenging activity of anthocyanin rich extracts from *Hibiscus rosa-sinensis*

Table 4 IC_{50} values of anthocyanin extracts obtained by NO antioxidant assay

IC_{50} (μg/mL)	
HRSAE	Ascorbic acid
62.25	35.28

Fig. 3 Correlation between in vitro antioxidant assay and **a** total phenolic content, **b** total flavonoid content and **c** total anthocyanin content

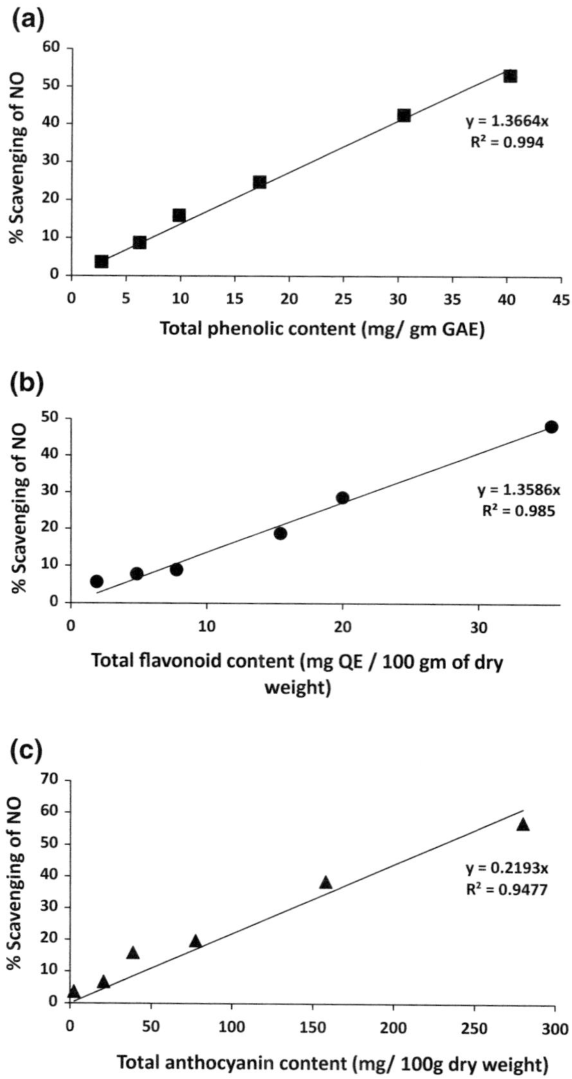

TAC; so these phytochemicals are believed to be a good source of antioxidants. Thus, this study highlighted *HRS* having antioxidant activity attributed to its anthocyanins, phenolics and flavonoids contents.

Table 5 Correlation between NO scavenging assay and with total phenolic, total flavonoid and anthocyanin content of the AEs (R^2, Regression coefficient and r, Pearson's coefficient)

	Total phenolic content	Total flavonoid content	Total anthocyanin content
NO scavenging assay	$R^2 = 0.994$ ($y = 1.3664x$) $r = 0.9980$	$R^2 = 0.985$ ($y = 1.3586x$) $r = 0.9931$	$R^2 = 0.9477$ ($y = 0.2193x$) $r = 0.9921$

Conclusions

The result of this study asserts the in vitro antioxidant capacity of extract of the flower of *HRS*, which is comparable to that of the standard. The phytochemicals like anthocyanins, polyphenols and flavonoids in the *HRS* flower makes it a pharmacologically effective antioxidant. The findings suggest that *HRS* could be a potential source of natural antioxidant that could have great importance as a therapeutic agent. Further investigation is necessary to establish an in vitro and in vivo correlation of antioxidant activities.

References

Andersen ØM, Markham K (2006) Anthocyanins. In: Flavonoids: chemistry, biochemistry, and applications, CRC Press, Boca Raton

Bhaskar A, Nithya V, Vidhya VG (2011) Phytochemical screening and in-vitro antioxidant activities of ethanolic extracts of *Hibiscus rosa-sinensis* L. Ann Bio Res 2:653–661

Bueno JM, Ramos-Escudero F, Sáez-Plaza P, Muñoz A, JoséNavas M, Asuero AG (2012) Analysis and antioxidant capacity of anthocyanin pigments. Part I: general considerations concerning polyphenols and flavonoids. Cr Rev Anal Chem 42:102–125

Chia LS, Goh NK, Chia TF, Brouillard R, Jin-Ming Konga (2003) Analysis and biological activities of anthocyanins. Phytochemistry 64:923–933

Faten R, Ghaffar A, El-Elaimy I (2012) In-vitro, antioxidant and scavenging activities of *Hibiscus rosa-sinensis* crude extract. J App Pharm Sci. 2:51–58

Giusti MM, Wrolstad RE (2001) Characterization and measurement of anthocyanins by UV-visible spectroscopy. In: Current protocols in food analytical chemistry, Wiley, NY, pp 1–13

Harman D (1956) Ageing: a theory based on free radical and radiation chemistry. J Gerontol 11:298–300

Halliwell B, Gutteridge JMC (1997) Free radicals in biology and medicine. Oxford University Press, Oxford

Kong JM, Chia LS, Goh NK, Chia TF, Brouillard R (2003) Analysis and biological activities of anthocyanins. Phytochemistry 64:923–933

Kumar A, Singh A (2012) Review on *Hibiscus rosa-sinensis*. Int J Res Pharm Biomed Sci 3:2229–3701

Lee J, Kim H, Kim J, Jang Y (2007) Antioxidant property of an ethanol extract of the stem of Opuntia ficus-indica var. Saboten. J Agric Food Chem 50:6490–6496

Liu X, Zhao M, Wang J, Yang B, Jiang Y (2008) Antioxidant activity of methanolic extract of emblica fruit (*Phyllanthus emblica* L.) from six regions in China. J Food Compos Anal 21:219–228

Moncada A, Palmer RMJ, Higgs EA (1991) Nitric oxide: physiology, pathology and pharmacology. Pharm Rev 43:109–142

Patel A, Patel A, Patel NM (2010) Determination of polyphenols and free radical scavenging activity of *Tephrosia purpurea* L. leaves (Leguminosae). Pharmacog Res 2:152–158

Prasad MP (2014) In vitro phytochemical analysis and antioxidant studies of *Hibiscus* species. Int J Pure App Bio 2:83–88

Singleton VL, Rossi JA (1965) Colorimetry of total phenolics with phosphomolybdic-phosphotungstic acid reagents. Amm J Enol Viti 16:144–158

Soto-Vaca A, Gutierrez A, Losso JN, Xu Z, Finley JW (2012) Evolution of phenolic compounds from colour and flavour problems to health benefits. J Agric Food Chem 60:6658–6677

Thaipong K, Boonprakob U, Crosby K, Cisneros-Zevallos L, Byrne DH (2006) Comparison of ABTS, DPPH, FRAP, and ORAC assays for estimating antioxidant activity from guava fruit extracts. J Food Comp Anal 19:669–675

Tiwari P, Kumar B, Kaur M, Kaur G, Kaur H (2011) Phytochemical screening and extraction: a review. Int Pharm Sci 1:98–106

Wang C, Wang J, Lin W, Chu C, Chou F, Tseng T (2000) Protective effect of Hibiscus anthocyanins against tert-butyl hydroperoxide-induced hepatic toxicity in rats. Food Chem Toxicol 38:411–416

Study of Biochemical Changes on Freeze-Dried and Conventionally Dried White Button Mushroom as a Sustainable Method of Food Preservation

Pinki Pal, A.K. Singh, Dipti Kumari, Rahul Rahul, J.P. Pandey and Gautam Sen

Abstract

The conventional biotechnology and the advanced agricultural practices may not be sufficient to meet the rising and expected high protein demands. Presently mushroom has occupied a high status because of its nutritive value, i.e. it is rich in protein and fibre along with several vitamins and minerals contents, it contains virtually no fat and cholesterol. However, preservation of mushroom during its storage and distribution has always been a challenge. Cryogenic freezing of food material is the only valuable option for its preservation. The principle behind the freeze drying involves the triple point of water. It occurs at 0 °C temperature and 4.7 mm of Hg pressure. Dehydration of food material has taken place by sublimation process. A comparative study on freeze-dried and other conventional dried mushrooms has been performed. White button mushrooms were preserved by hot air oven method and freeze-drying method. Nutritional parameters like moisture, protein, crude fat, reducing sugars, ash content, crude fibre, ascorbic acid content and sensory evaluation like texture, flavour, aroma, shape, rehydration ratio, particle nature and storage stability of samples dried by different methods have been studied by different chemical methods. The freeze-drying technique has been found excellent for mushroom dehydration, since the quality of product is found to be the best in terms of its colour, flavour, texture and nutritive value. Therefore, this method is found to be useful for preservation of mushrooms.

P. Pal (✉) · D. Kumari · R. Rahul · J.P. Pandey · G. Sen
Department of Chemistry, Birla Institute of Technology, Mesra, Ranchi 835215, Jharkhand, India
e-mail: pinkipalhbti@gmail.com

A.K. Singh
Department of Biochemical Engineering and Food Technology, Harcourt Butler Technological Institute, Kanpur 208002, Uttar Pradesh, India

K. Mukhopadhyay et al. (eds.), *Applications of Biotechnology for Sustainable Development*, DOI 10.1007/978-981-10-5538-6_18

Keywords
White button mushroom · Freeze drying · Oven drying · Nutritional value · Triple point of water

Introduction

Mushroom, saprophytic macroscopic fungi, which are devoid of chlorophyll must obtain their carbohydrates from decomposed organic matter. There are around 38,000 mushroom varieties of which about 100 are considered edible, most popular one are the *Agaricus bisporus* (white button), *Lentinus edodes* (Shiitake) pleurotus species like *P. ostreatus* (Oyster), *P. sajor-caju*, *Flammulina velutipis*, *Auricularia polytrica*, etc.

Preservation means increasing the time for spoilage of food materials as much as possible. In case of mushroom, some techniques used for its preservation are blanching, steeping, vacuum cooling, drying (high temperature and freeze drying), osmotic dehydration, canning, freezing, irradiation (Drouzas and Schubert 1996; Prabhanjan et al. 1995) and control atmospheric storage; but these methods results in brown discolouration, loss of flavour and aroma, change in texture and appearance and loss of nutritive value. However, this magnitude of over and ever decreasing quality of mushroom during preservation warrants exploration of unconventional methods of preservation of mushrooms to make it a valuable supplement of protein and to increase the shelf life of food.

Freeze drying is an emerging technology for drying of mushrooms, as it overcomes, in whole or in part, all these undesirable characteristics. The freeze-dried mushrooms have the best quality, among the mushrooms dehydrated by other methods (Hammami and Rene 1997; Rey and Bastein 1962). The aroma retention by freeze-dried mushrooms is also reported to be excellent (Ray and Bastein 1962).

In the study reported here, white button mushroom has been successfully preserved by freeze drying and oven drying method. Comparative study on nutritional analysis and sensory characteristics of mushroom persevered by different methods has been made.

Materials and Methods

Materials: Fresh mushroom samples were purchased from market. All the chemicals used were analytical grade reagent. These were used without further purification.

Preservation by Oven Drying (Conventional Method)

Fresh mushrooms were blanched with steam/hot water for short period (2 min) to inactivate the enzymes, catalase and peroxidise (Fang et al. 1971). Adequacy of blanching was tested by catalase test and peroxidase test. Blanched mushrooms were treated with Potassium meta bi Sulphite (KMS) solution (Fang et al. 1971). KMS treated mushrooms were dried in hot air oven at about 100 °C. After complete drying (Sugana et al. 1995), they were cooled in desiccator and weighed. It was then packed in air tight polyethylene bags.

Preservation by Freeze Drying

Freeze drying is an advanced method of food preservation involving freezing by water removal through sublimation. Mushroom freeze-drying method (Hammami and Rene 1997; Rey and Bastein 1962) is summarized Fig. 1.

Freeze dryer (Alpha 2-4, Machine No. 100402, Martin Christ Germany) was used for dehydration of samples. The drying was done at 0 °C temperature under

Fig. 1 Flow chart of freeze drying

Harvesting (90% moisture content)

↓

Washing

↓

Blanching (in hot water/ steam)

↓

Steeping (KMS solution at 200 ppm SO_2)

↓

Slicing (using hand operated slicer)

↓

Loading (uniform loading on trays)

↓

Freezing (- 32°C temp., 4-6 hrs)

↓

Freeze drying (at 4.7 mm of Hg pressure)

↓

Freeze dried product (3% moisture content)

↓

Packing (Al. Foils/ polythene bags)

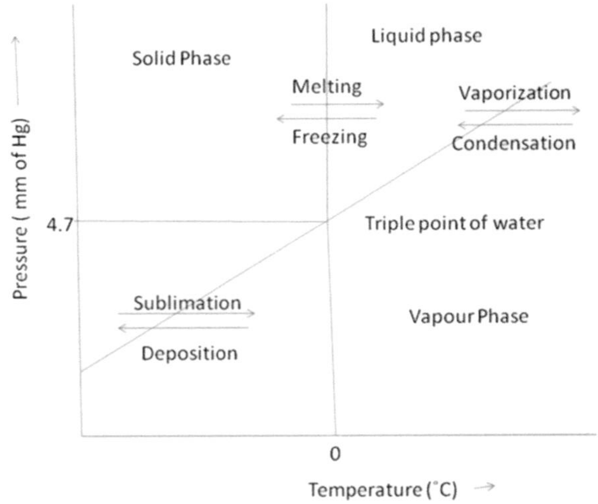

Fig. 2 Triple point of water

the pressure of 4.7 mm of mercury (triple point of water, i.e. when all three physical states of water exist simultaneously (Fig. 2)). After drying, the product was packed in Aluminium foils.

Analysis of Oven-Dried and Freeze-Dried Mushrooms

The dried mushroom samples were analysed for proximate constituents, dehydration ratio, rehydration ratio and sensory characteristics.

Proximate Constituents

The moisture content of dried mushroom samples were analysed by oven drying method (measuring the weight lost by samples due to evaporation of water). Estimation of reducing sugar was done by Dinitrosalicylic acid (DNS) method. Protein content was estimated by Biuret method/kjeldahl method. Crude fat of the samples was estimated by soxhlet extraction method using *n*-hexane as a solvent. Total ash was estimated by using muffle furnace, and ascorbic acid (Vitamin C) content was estimated by 2,6-Dichlorophenol-indophenol method (visual titration method). Crude fibre is the organic residue (cellulose together with a little lignin) which remains after the food sample has been treated under standardized condition with petroleum ether, boiling dil. Sulphuric acid, boiling dil. Sodium hydroxide solution and alcohol. Dehydration ratio and rehydration ratio was determined by gravimetric method. The results of analysis of all the chemical parameters are given in Table 1. The comparative data regarding quality of oven-dried and freeze-dried samples have been presented in Table 2.

Table 1 Proximate constituent of mushroom dried by different methods

S. No.	Proximate constituents	Fresh mushroom (%)	Oven-dried mushrooms (%)	Freeze-dried mushroom (%)
1.	Moisture	90.10	BDL	BDL
2.	Protein	03.79	30.18	34.10
3.	Crude fibre	00.12	–	06.07
4.	Reducing sugar	00.50	01.75	03.06
5.	Crude fat	00.20	03.70	4.00
6.	Total ash	00.68	09.80	07.20
7.	Carbohydrate	05.11	–	45.26
8.	Ascorbic acid	02.15	20.00	95.60[a]
9.	Dehydration ratio	–	15.8:1	12:1
10.	Rehydration ratio	–	1:2.45	1:7

BDL Below detectable limits
[a]mg/100gm

Table 2 Comparative study of quality parameters of oven-dried and freeze-dried mushrooms

S. No.	Characteristics	Oven-dried mushroom	Freeze-dried mushroom
1.	Colour	brown	Creamy white
2.	Aroma	Abnormal	Natural
3.	Texture	Hard	Soft and brittle
4.	Flavour	Abnormal	Almost natural
5.	Density of the product	Higher than original food	Lower than original food
6.	Particle nature	Solid	porous
7.	Rehydration	Slow and incomplete	Rapid and almost complete
8.	Texture after rehydration	Hard and rubbery	Very soft
9.	Shape after rehydration	Not original	Same as original
10.	Storage stability	Good	Excellent

Results and Discussions

Preservation by Oven Drying (Conventional Method)

From the results, the oven-dried mushrooms have lost its natural colour (Krokida et al. 1998) because of the activity of enzymes. Aroma and flavour are also lost. This may be, since, at higher temperature some volatiles are escaped. The density of the product is more than the fresh one as it consists of solid particles. Rehydration of oven-dried samples was very slow and incomplete because it possesses hard and rubbery texture. The oven-dried mushroom has also lost its original shape and appearance.

The major undesirable changes are mainly due to the following:

1. Pronounced shrinkage of solids.
2. Migration of dissolved constituents to the surface when drying solid.
3. Extensive denaturation of protein.
4. Case-hardening—The formation of relatively hard, impervious layer at the surface of a solid is caused by one or more of the first three changes. This impervious layer slows rates of both dehydration and reconstitution.
5. Undesirable chemical reaction.
6. Excessive loss of desirable volatile constituents.
7. Difficulty of rehydration as a result of one or more of the above changes.
   Besides these undesirable effects, oven-dried mushroom has low protein and fat content. From Table it can be observed that the ascorbic acid was almost completely lost in oven drying method. Oven-dried samples have been shown good storage stability.

Preservation by Freeze Drying

It is evident from Table 2 that the freeze-dried mushrooms retain its original colour, flavour and texture because blanching inactivates the enzymes responsible for discolouration. At low temperature, flavour constituting substances cannot volatilize. Formation of hard, impervious solid is not there, since in freeze-drying water is directly converted to gaseous phase without being converted into liquid phase (sublimation) (Genin et al. 1996; Guildner et al. 1976). The original shape and texture remains preserved. The sample becomes porous and soft which has density lower than the fresh food and it rehydrate rapidly and almost completely. Storage stability of these samples was excellent. The protein and fat content of freeze-dried mushrooms have been found higher in comparison to oven-dried samples (Hayes 1972).

Conclusions

The quality and nutritional content of freeze-dried mushrooms was found to be much desirable in comparison to oven-dried mushrooms. Superiority of the product was found to be the best in terms of its colour, flavour, texture, shape, rehydration property and storage stability, etc. However, freeze drying is an expensive process and consequently, the operation has to be carried out with in optimum condition for a maximum quality preservation in order to compensate its cost. A lot of careful planned research and development programme is needed to develop this technology for the preservation of many other kinds of agricultural residues.

Acknowledgements The authors greatly acknowledge the financial support from Department of Science and Technology (DST), New Delhi, India in form of a research grant (sanction order No. SR/WOS- A/ET- 13/2014) to carry out the reported investigation. The authors also deeply acknowledge the support of Department of Chemistry, Birla Institute of Technology, Mesra, Ranchi for their active support.

References

Drouzas AE, Schubert H (1996) Microwave application in vacuum drying of fruits. J Food Eng 28:203–209

Fang TT, Footrakul P, Luh VS (1971) Effects of blanching, chemical treatments and freezing methods on quality of freeze-dried mushrooms. J Food Sci 36(4):1044–1048

Genin N, Rene F, Corrieu GC (1996) A method for on-line determination of residual water content and sublimation end point during freeze drying. Chem Eng Process 35:255–263

Guildner LA, Johnson DP, Johns FE (1976) Vapour pressure of water at its triple point. J Res Natl Bur Stand-A Phys Chem 80A(3):505–521

Hammami C, Rene F (1997) Determination of freeze drying process variable for strawberries. J Food Eng 32:133–154

Hayes WA (1972) Nutritional factor in relation to mushroom production. Mushroom Sci 8:663–674

Krokida MK, Tsami E, Marculis ZB (1998) Kinetics on colour changes during drying of some fruits and vegetables. Dry Technol 16:667–685

Prabhanjan DG, Ramaswamy HS, Raghvan GSV (1995) Microwave- assisted convection air drying of the thin layer carrots. J Food Eng 25:283–293

Rey LR, Bastein MC (1962) Freeze-drying of foods. Natl Acad Sci 25–42

Sugana S, Usha M, Sreenarayanan VV, Raghupathy R, Gothandapani L (1995) Dehydration of mushroom by sun drying, thin layer drying, fluidized bed drying and solar cabinet drying. J Food Sci Technol 32:284–288

In Silico Modelling of Hepatocellular Carcinoma Linked PARP-1 Protein and Screening of Potential Inhibitors

Santosh Kumar Jha, Hare Ram Singh, Rati Kumari Sinha and Pragya Prakash

Abstract

Hepatocellular carcinoma (HCC) is one among the hard-treating cancer with a high mortality rate and there is an urgent demand of more effective therapeutics for it. So many complications and side effects are associated with present available treatment modalities. There is a scarcity of drug molecules for HCC and most of them are (viz. Sorafenib) under clinical trials. Few proteins are reported which are linked with hepatocellular carcinoma. The HCC-linked protein Poly[ADP-ribose]polymerase-1 (PARP-1) is identified as a druggable target. We have used a computational method to find novel inhibitor as potential drug candidate molecule to treat HCC. Ninety-one conformations of PARP-1 inhibitors were designed for standard precision in silico docking. The different physical parameter scores, viz. docking score and glide score, were used to identify the suitable inhibitors. The best-docked inhibitors were further characterized by ADMET analysis, including Lipinski's rule of five, Jorgensen's rule of three, blood–brain barrier penetration index, skin permeability index, human intestinal absorption index and oral absorption index. $C_{15}H_{19}N_3O_2^+$ (PJ3410) was identified as the best inhibitor of PARP-1 among all the selected 91 inhibitors conformers with best ADMET properties. PJ3410 may be a better drug molecule for the inhibition of HCC-linked PARP-1 protein. Present investigation reveals the importance of structure-based computational drug design methodology which is time and cost-effective.

Keywords

ADMET · HCC · Jorgensen's rule · Lead molecule · Lipinski's rule · PARP-1

S.K. Jha (✉) · H.R. Singh · R.K. Sinha · P. Prakash
Department of Bio-Engineering, Birla Institute of Technology, Mesra, Ranchi 835215, Jharkhand, India
e-mail: skjha@bitmesra.ac.in

K. Mukhopadhyay et al. (eds.), *Applications of Biotechnology for Sustainable Development*, DOI 10.1007/978-981-10-5538-6_19

Introduction

Hepatocellular carcinoma is one among the most common types of primary malignancy of liver, followed by cholangiocarcinoma. In most cases, viral infections (hepatitis B or C) or cirrhosis may result in the HCC. Although most HCCs develop in the background of chronic liver disease, some may occur in normal liver and usually correspond to a specific type, including fibrolamellar HCC mostly encountered in young populations, or malignant transformations of hepatocellular adenomas (Paradis 2013). The main risk factors that cause HCC are alcoholism, Hepatitis B or C, Aflatoxin consumption, Wilson's disease, etc. Patients with HCC have one or more clinical features including right upper quadrant pain, weight loss, worsening of liver enzyme, acute abdominal catastrophe, intra-abdominal bleeding, hepatic bruit, etc. (El-Serag and Mason 2003).

It was reported that poly(ADP-ribose) polymerase-l (PARP-l) has a crucial role in repair after the DNA damage, cell death and cell proliferation. It is also associated with the stabilization of the genome. Genetic ablation or pharmacological inhibition of Poly[ADP-ribose]polymerase-1 (PARP-1) had a promising outcome in the chemotherapy of cancer. Huang et al. (2008) reported the PJ34 [*N*-(6-oxo-5, 6-dihydrophenanthridin-2-yl)-*N*, *N*-dimethylacetamide HCl] as an effective inhibitor of PARP-l. The pharmacological action of PJ34 was associated with the reduction in VEGF-induced proliferation and migration as well as tube formation of human umbilical vein endothelial cells. It was a dose-dependent phenomenon when studied in vitro, suggesting that the treatment of carcinoma with PJ34 may impart an additional benefit in various cancers by decreasing the process of angiogenesis (Rajesh et al. 2006). It has also been studied that this inhibitor was able to suppress cell growth of liver cancer cell lines in vitro.

In this chapter, structure-based drug designing (SBDD) methodology is adopted to sort out the potential inhibitors (Mandal et al. 2009). Several modifications in the structure of selected ligands were made to analyse the consequences and effectiveness to which structural variations contribute to ligand-PARP-1 binding which may result in inhibition. The designing and modification of ligand molecules were carried on the basis of structural information of the active site of the PARP-1, so that designed inhibitors can effectively bind (dock) with a higher affinity with target protein and may inhibit the function of it. The ligand having best binding pose along with minimum binding energy or maximum stabilization may be proposed as potential candidate molecules to develop the drug against the HCC.

Methodology

Sequence retrieval and analysis: The amino acid sequence of PARP-1(Human) (gi|156523968, accession no. NP_001609) protein in FASTA format was retrieved from National Centre for Biotechnology Information (NCBI). It had 1014 amino acid residues. BlastP module of basic local alignment search tool (BLAST) (www.blast.ncbi.nlm.nih.gov) using Smith–Waterman algorithm was used to analyse the sequence of PARP-1. Based on minimum E-value, maximum identity and minimum resolution, seven different protein templates were selected for multiple-template modelling.

Computational modelling of structure of PARP-1 Protein: The 3D structure of the PARP-1 protein was predicted with MODELLER9.11 (Eswar et al. 2006) tool. The python script files were used to execute the suitable modeller commands. Since the single suitable template was unable to be found by BLAST result for PARP 1, multiple templates based structural modelling was performed by using seven templates. The templates with PDB ID 4DQY C, 2RIQ A, 2CS2 A, 2COK A, 2CR9 A, 2RD6 A and 2L30 A at resolutions 3.25, 1.70, 99.9, 99.9, 99.9, 2.30 and 99.9 Å, respectively, were selected for the structure prediction by borrowing their geometrical coordinates from RCSB protein data bank (www.rcsb.org). The structure prediction by using multiple-template method is more accurate because it can align a single protein sequence simultaneously to multiple templates. The improvement in pairwise sequence-template alignment accuracy as well as generated models with better quality than single-template models could be obtained by multiple-template method (Peng and Xu 2011).

Model evaluation and validation: The predicted model of PARP-1 was evaluated and validated by complete stereochemical and geometrical analysis using discrete optimized potential energy (DOPE) score, Procheck (www.ebi.ac.uk/thornton-srv), Molpdf, ERRAT score (http://nihserver.mbi.ucla.edu/ERRATv2) and What If (http://swift.cmbi.ru.nl/whatif). The feasibility of structural stability of proposed model was judged by Ramachandran plot using an online server SAVes (http://services.mbi.ucla.edu/SAVES). Refined and validated model of PARP-1 protein was submitted to protein model database (http://mi.caspur.it/PMDB/) (Table 1).

Table 1 Ramachandran plot statistics for modelled PARP-1 protein

Residues in most favoured region [A, B, L]	776	86.9%
Residues in additional allowed regions [a, b, l, p]	105	11.7%
Residues in generously allowed regions [−a, −b, −l, −p]	10	1.1%
Residues in disallowed regions	6	0.7%
Number of non-glycine and non-proline residues	897	100%
Number of end-residues (excluding Gly. and Pro)	2	
Number of glycine residues (shown as triangles)	71	
Number of proline residues	44	
Total number of residues	1014	

Standard Precision Screening of Inhibitors and Molecular Docking

Schrodinger product suits were used for the preparation ligand, docking study and analysis of ligand–receptor interactions by hierarchical search protocol-based algorithms. The energy minimization of a flexible ligand in the field of the Coulomb and Vander Waals potential of the protein is the final step in this protocol.

Ligand preparation: Literature survey and chemical databases were used to collect the basic information about various ligands or inhibitors of PARP-1. The analogues of *N*-(6-oxo-5, 6-dihydrophenanthridin-2-yl)-*N*, *N*-dimethylacetamide HCl were designed as the potential inhibitors. Initially, 20 different structural analogues were considered for molecular docking. The designed structures of the different inhibitors were drawn in Maestro (Schrodinger). LigPrep module of Schrodinger product suits was used for final ligand preparation as well as clean up the structure and assigns the stable conformations of each ligand structure in three-dimensional spaces. The one or two stereoisomer of each analogue was also generated. The details of all the designed structures and isomers of selected inhibitors were saved automatically in the maestro project table.

The receptor grid was generated at charge scale factor of 1.0 and Vander Waal radius factor scale of 1.0 by selecting residues present at active sites. The docking was performed by glide module. The residues present at active sites were identified on the basis of literature survey and using ligand interaction menu of Maestro.

Prediction of Absorption Distribution Metabolism Excretion Toxicity (ADMET analysis)

QikProp module of Maestro was used for in silico characterizations of selected ligands in normal mode. It is a fast, robust and accurate prediction programme for absorption, distribution, metabolism, excretion and toxicity (ADMET). This module can predict physically significant descriptors and pharmaceutically relevant properties of ligand molecules, either individually or in groups. These characterized inhibitors (ligands) may be considered as potential lead molecules. Finally, the Lipinski's rule of five and Jorgensen's rule of three were used to assess the qualities of lead molecules.

Result and Discussion

Structural Modelling of PARP-1

Multiple templates based homology modelling is an accurate method for structure prediction. The degree of sequence identity between target and template as well as the resolution of the templates decides the quality of homology modelling.

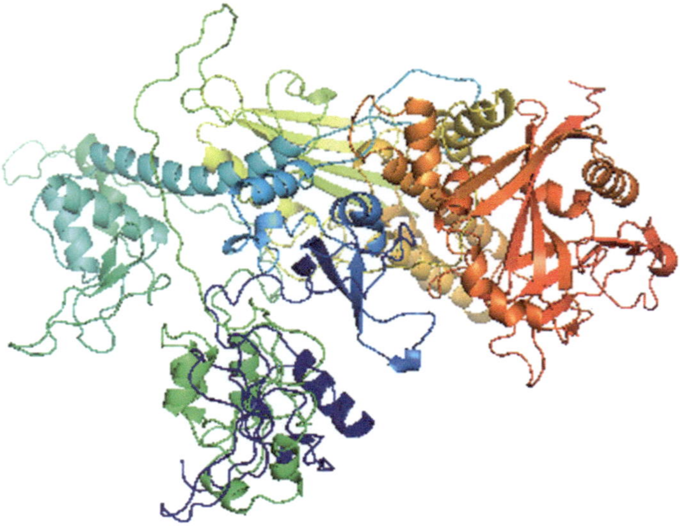

Fig. 1 Modelled structure of PARP-1

Table 2 Docking and related energies score for best five inhibitors of PARP-1 protein

S. No.	Molecular formulae of inhibitors	Docking score	g-ecoul	g-evdw	Glide lipo	g-energy	Potential energy
1.	$C_{15}H_{19}N_3O_2^{2+}$	−8.20238	−23.2319	−21.9164	−1.08063	−45.1483	101.1816
2.	$C_{16}H_{21}N_3O_2^{2+}$	−8.02587	−23.4258	−24.6478	−1.19845	−48.0736	108.7135
3.	$C_{17}H_{24}N_3O_2^{3+}$	−7.34407	−23.4237	−18.8165	−0.69197	−43.7664	118.5646
4.	$C_{15}H_{16}N_3O_2^{+}$	−6.87126	−10.7271	−23.9834	−1.07443	−34.7106	72.17025
5.	$C_{16}H_{18}N_3O_2^{+}$	−6.59683	−10.8966	−30.3691	−1.64818	−41.2658	89.34424

Modeller9.11 generated the total 20 different models of PARP-1 (Fig. 1). The best model (1fdx.B99990013.pdb) among all was identified by considering minimum DOPE score (−85,526.46094), minimum molpdf (32,644.05469) and GA341 scores (1). This model has all other stereochemical parameters within the acceptable limit. The stereochemical parameters, viz. RMS Z-Score for bond angle (1.121), bond length (0.89), omega angle restraints (1.31), improper dihedral distribution (0.782), side-chain planarity (1.420) and outside distribution values (1.07), also indicate the good quality of model prediction. The modelled PARP-1 protein had acceptable physiochemical environments as well as the overall quality factor of 93.14. The Ramachandran plot was used to analyse the backbone phi (φ) and psi (ψ) dihedral angles. The plot indicates the fairly good percentage of residues (86.7%) were in the most favoured region, while the residues in additional allowed and generously allowed regions were 11.7 and 1.1%, respectively. Only 0.7% of

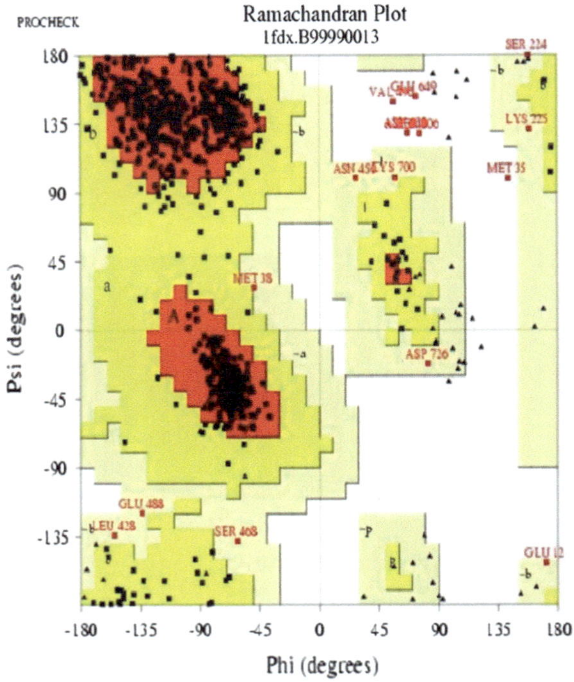

Fig. 2 Ramachandran plot of modelled structure of PARP-1

residues fall under disallowed region (Table 2). The low-energy zones were considered as allowed regions in the plot. The highly favoured region on the plot is represented by red colour, while the additional allowed regions and generously allowed region are in yellow and light yellow colour, respectively (Fig. 2).

Ligand Docking

The three-dimensional structure of the receptor of target protein (PARP-1) obtained from modelling was used in the docking study. All the 91 selected ligands including the structural and stereoisomeric analogue of identified inhibitor were used to dock after energy minimization and generating the grid as described earlier. The docking score was used to preliminary identify better ligands. The secondary screening of ligands was further carried out by analyzing the degree of their hydrophobicity, extent of Columbic interaction, level of polar interactions, internal torsional energy and rotatable bonds (Table 2). The arrangement of molecules and their position decided the potential energy of a molecule. The stable complex can be formed by

Table 3 Specification of best five inhibitors

S. No.	IUPAC name of inhibitors	Molecular formulae	Molecular weight	Structure
1.	PJ3410	$C_{15}H_{19}N_3O_2^{2+}$	273.33	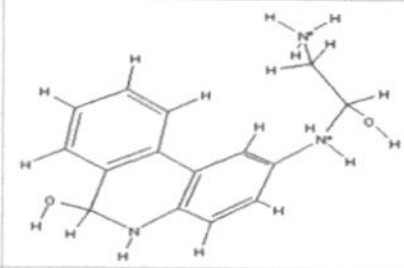
2.	PJ3411	$C_{16}H_{21}N_3O_2^{2+}$	287.36	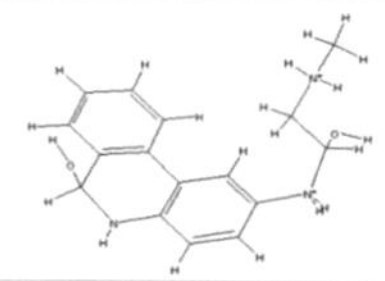
3.	PJ349	$C_{17}H_{24}N_3O_2^{3+}$	302.39	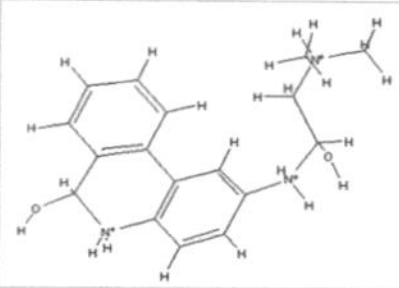
4.	PJ346	$C_{15}H_{16}N_3O_2^{+}$	270.31	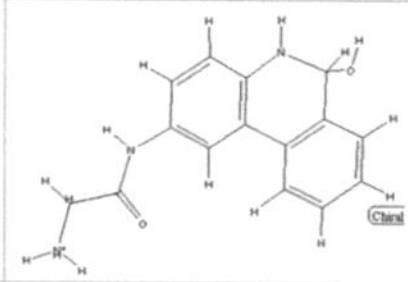
5.	PJ3412	$C_{16}H_{18}N_3O_2^{+}$	284.33	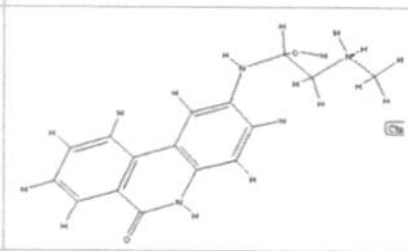

molecules with low potential energy. The binding affinity of ligand and receptor can be estimated by their docking score after their interaction in docking pose. The ligands [PJ3410, PJ3411, PJ3419, PJ349 and PJ3417] have comparatively similar docking score among all the selected and designed inhibitors. PJ3410 shows minimum docking score of −8.20238. Based on docking score, PJ3410 has appreciably better binding affinity in comparison to other inhibitors. On this basis, newly designed molecule PJ3410 may be more potent and selective towards the inhibition of PARP-1. Few other ligands had comparatively same docking features, which may also be used as potential lead molecules (Tables 2 and 3).

The amino acid residues present at active site or around the ligands in the receptor protein were TRY(8890), TYR(896), HIS(862), GLY(894), ILE(879), ALA(880), ARG(878), LEU(877), ASP(770), ASP(766), ASN(767), ARG(865), SER(864), ASN(868), GLU(763), HIS(909), TYR(689) and TYR(907) (Fig. 3). The interaction between ligand and PARP-1 reveals the Van der Waal and Columbic nature of association between them with an energy value of −21.9164 and −23.2319, respectively, for best-docked ligand PJ3410. The screened ligands have a tendency to bond with the hydrogen bonds with a target which has stronger

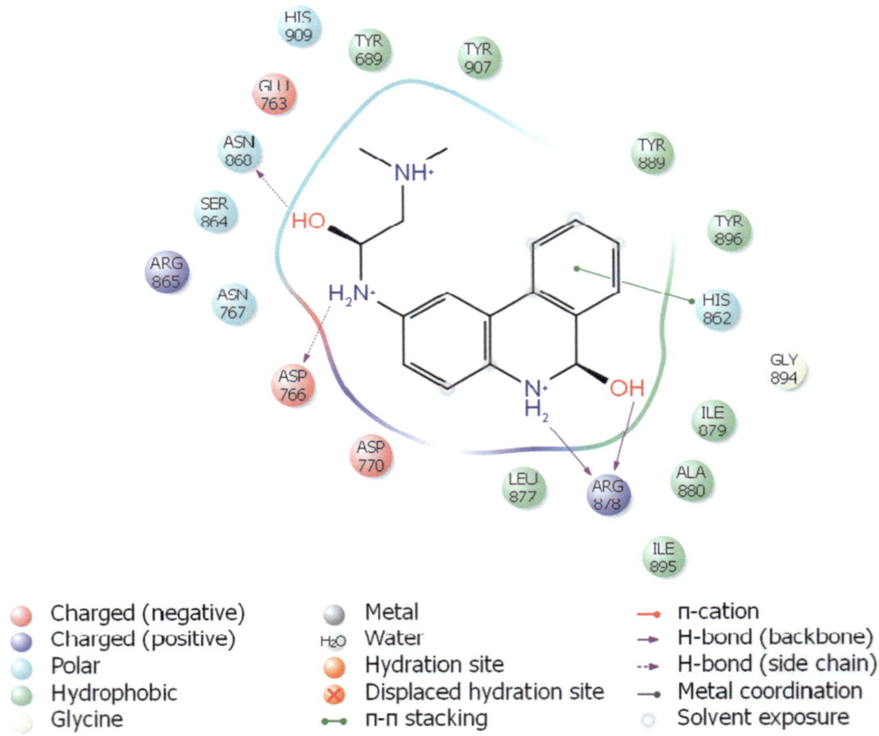

Fig. 3 Amino acid residues present in the active site of the PARP-1 protein

interaction than Van der Waal and Columbic forces. The major amino acids, viz. ASN(868), ASP(766) and ARG(878), of PARP-1 are involved in the establishment of hydrogen bonds with the ligand (Figs. 4 and 5). The possibilities of establishment of hydrophobic interaction with the hydrocarbon moiety of the ligands are clear due to the presence of ligand in the hydrophobic pocket at the time of interaction (Fig. 6). The ligand PJ3410 having lowest (−1.08063) hydrophobic interaction energy reflects the overall good quality parameters. Hence, here we may propose PJ3410 as a better lead molecule to develop the drug against the HCC.

ADMET Analysis

The top ten inhibitors of PARP-1 were further subjected to ADMET analysis to access their pharmacological properties and treating them as potential drug candidates. A good drug candidate should satisfy the Lipinski's Rule of five and Jorgensen's rule of three. After calculating ADMET, it is concluded that PJ3410 has

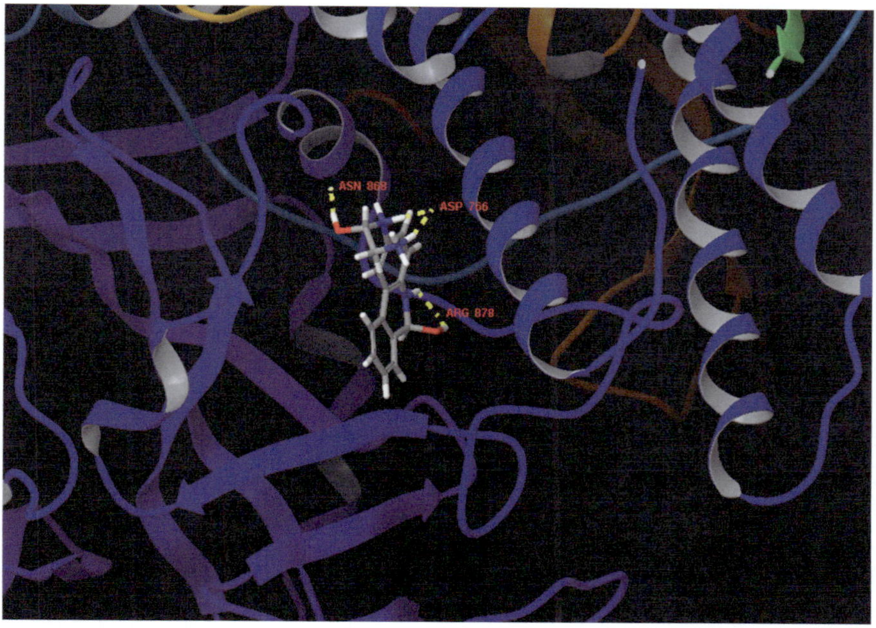

Fig. 4 Protein–ligand interaction showing receptor–ligand hydrogen bond and intra-ligand contacts

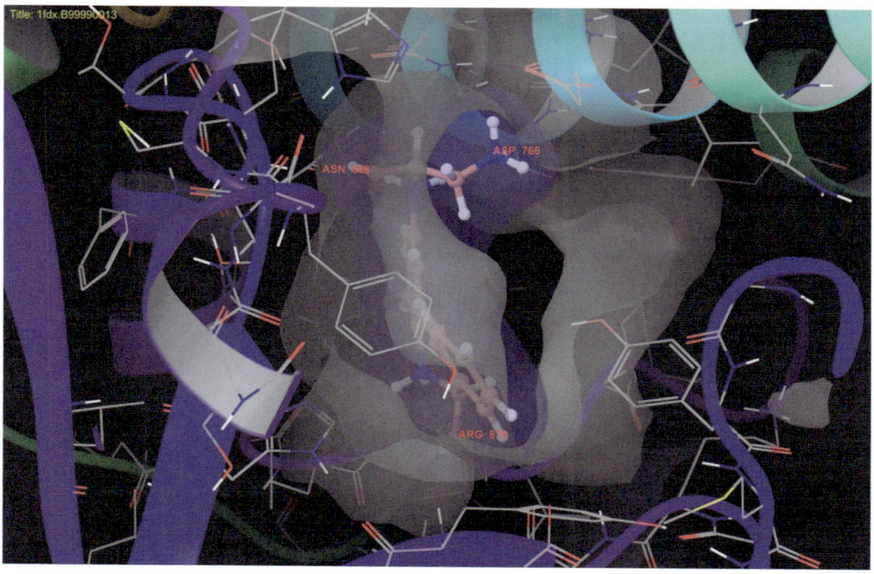

Fig. 5 Ligand bound to the PARP-1 protein through hydrogen bonds with ASN(868), ASP(766) and ARG(878)

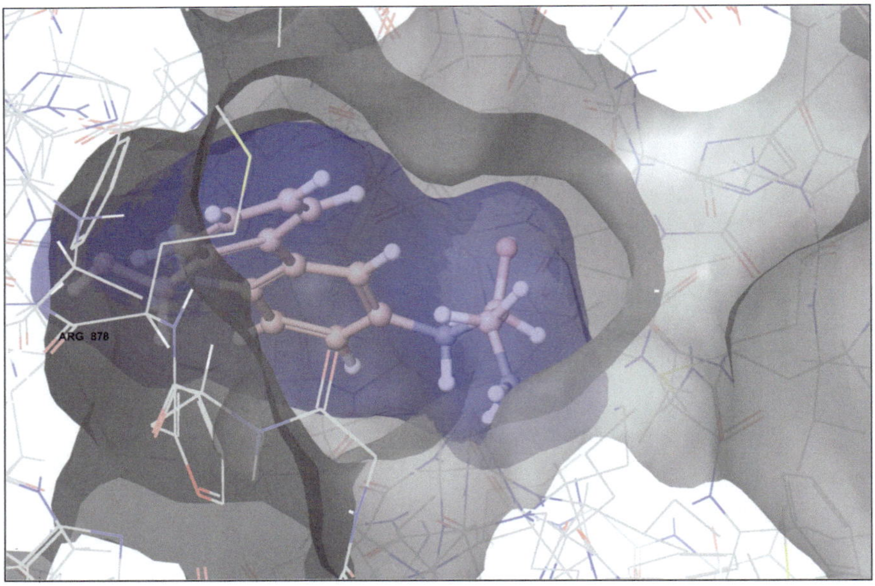

Fig. 6 Ligand bound to the PARP-1 protein through hydrophobic interaction with hydrophobic moieties

obeyed the rule of five and rule of three at the level of satisfaction. It was estimated that all the top ten inhibitors have more than 80% of human intestinal absorption rate which was within the acceptable range.

The blood–brain barrier (BBB) penetration value predicted by QikProp for screening inhibitors was within the satisfactory limit of acceptance. The ease to cross the blood and brain barrier by the drug molecules can be accessed by BBB penetration index. As a thumb rule, good quality of inhibitor or drug molecules should have the capacity to cross the barrier. The CNS active drug molecules must pass across it and CNS-inactive drug molecules must not pass across to avoid CNS-associated side effects. QPlogBB denotes blood partition coefficient of selected inhibitors, and all of them have acceptable ranges between −3.0 and −2.

QPlogKp value gives the information about skin permeability rate of drug that exposed with the skin either accidentally or by design. In present study, QPlogKp value was found in the range of −8.0 to −1.0. In this study, PJ3410 inhibitor has satisfied all the above ADME properties. So it may be a potential drug candidate for the treatment of HCC. The summary of ADMET properties of best five screened inhibitors is given in Table 4.

Table 4 Comparison of ADMET properties of selected ligands

ADMET properties	Name of inhibitors				
	$C_{15}H_{19}N_3O_2^{2+}$	$C_{16}H_{21}N_3O_2^{2+}$	$C_{17}H_{24}N_3O_2^{3+}$	$C_{15}H_{16}N_3O_2^{+}$	$C_{16}H_{18}N_3O_2^{+}$
CNS	+2	+2	+2	+1	+1
Donor HB	6	5	4	5	4
Acceptor HB	6.4	6.9	7.4	6.2	6.7
QPP Caco (nm/s)	39.586	96.232	175.997	49.44	81.033
QPlogBB	−1.086	−0.732	−0.489	−0.94	−0.836
QPP MDCK	16.686	43.583	83.697	21.217	36.193
QPlogKp	−5.629	−4.898	−4.392	−5.571	−5.086
QPlogS	−0.982	−1.409	−1.691	−1.598	−1.959
QPlogPo/w	0.139	0.505	1.957	0.167	0.672
SASA	508.769	534.075	52.749	519.01	552.22
Human oral absorption	2	3	3	2	3
Percent human oral absorption	41.764	65.4	72.736	58.249	65.041
Mol_MW	271.318	285.345	299.372	269.302	283.329

Conclusion

Present work on the protein PARP-1(Poly[ADP-Ribose]polymerase-1) demonstrated that on the basis of templates, model of proposed protein was made and validated. Multiple-template homology modelling generates a better 3D model than single-template modelling. Docking software Schrodinger was used for designing inhibitor and protein–ligand docking. It was found that the screened ligands interact with the target site. The lowest GLIDE score was selected by considering different poses of ligand in the active site. The lowest GLIDE energy score indicates high affinity and specificity between the target and the ligand. The binding of protein and ligand is based on the interaction at the active site. These interactions mainly include hydrogen bonds and hydrophobic interactions. The greater the number of bonds the higher the binding affinity. On the basis of docking score, it is concluded that PJ3410 shows minimum docking score than other inhibitors and has not violated the Lipinski's rule of five. Therefore, we can conclude that PJ3410 can be considered as a good drug candidate for PARP-1 inhibition. After the in silico study, validation through the wet lab is necessary to arrive at a definite conclusion.

References

El-Serag HB, Mason AC (2003) Risk factors for the rising rates of primary liver cancer in the United States. Arch Intern Med 160:3227–3230

Eswar N, Webb B, Marti-Renom MA, Madhusudhan MS, Eramian D, Shen MY, Pieper U, Sali A (2006) Comparative protein structure modeling using Modeller. Curr Protoc Bioinform. Chapter 5 (unit 5.6)

Huang SH, Xiong M, Chen XP, Xiao ZY, Zhao YF, Huang ZY (2008) PJ34, an inhibitor of PARP-1, suppresses cell growth and enhances the suppressive effects of cisplatin in liver cancer cells. Oncol Rep 20:567–572

Mandal S, Moudgil M, Mandal SK (2009) Rational drug design. Eur J Pharmacol 625(1–3): 90–100

Paradis V (2013) Histopathology of hepatocellular carcinoma. Recent Results Cancer Res 190: 21–32 (Cancer Research)

Peng J, Xu J (2011) A multiple-template approach to protein threading. Proteins 79(6):1930–1939

Rajesh M, Mukhopadhyay P, Batkai S, Godlewski G, Hasko G, Liaudet L, Pacher P (2006) Pharmacological inhibition of poly (ADP-ribose) polymerase inhibits angiogenesis. Biochem Biophys Res Commun 350:352–357

Ferulic Acid Decarboxylase from *Bacillus cereus* SAS-3006: Purification and Properties

Shashank Mishra, Neha Panjiar, Ashish Sachan, Ambarish Sharan Vidyarthi and Shashwati Ghosh Sachan

Abstract

The cell-free extract of *Bacillus cereus* SAS-3006 strain grown on the medium containing ferulic acid was used as a source of the enzyme ferulic acid decarboxylase. This key enzyme, playing a vital role in the conversion of ferulic acid, was purified to homogeneity by anion-exchange chromatography from *Bacillus cereus* SAS-3006. The optimal pH and temperature for enzyme activity were found to be 6.5 and 37 °C, respectively. Kinetic studies indicated the K_m and $V_{\max}$ values of the purified ferulic acid decarboxylase were 0.0118 mM and 0.333 U, respectively. The enzyme activity was not inhibited by any of the studied metal ions viz. Fe^{2+}, Fe^{3+}, Ca^{2+}, Mg^{2+}, Mn^{2+}, and Zn^{2+} at 2.0 mM of concentration. The enzyme was found to be activated and expressed inside the cells adequately by the substrate ferulic acid.

Keywords

Ferulic acid decarboxylase · Ferulic acid · 4-vinylguaiacol · *Bacillus cereus* SAS-3006

S. Mishra · N. Panjiar · A. Sachan · S.G. Sachan (✉)
Department of Bio-Engineering, Birla Institute of Technology, Mesra, Ranchi 835215, Jharkhand, India
e-mail: ssachan@bitmesra.ac.in

S. Mishra · A.S. Vidyarthi
Department of Biotechnology, Institute of Engineering & Technology, Sitapur Road, Lucknow 226021, Uttar Pradesh, India

K. Mukhopadhyay et al. (eds.), *Applications of Biotechnology for Sustainable Development*, DOI 10.1007/978-981-10-5538-6_20

Introduction

Aromatic compounds play an important role in food, flavour, fragrance, and chemical industries. They are conventionally manufactured from petroleum by chemical synthesis, which currently faces lack of resources and energy with environmental pollution (Giedraityte and Kalediene 2014). Bioconversion and biodegradation of plant-related phenylpropanoid compounds such as ferulic acid are very significant because they are natural renewable resources and can be used for the production of desired natural compounds (Mishra et al. 2014a). Thus, biotransformation of ferulic acid has always been a burning topic, because it is a natural renewable resource and biodegradation processes are nature friendly. Ferulic acid, a derivative of 4-hydroxycinnamic acid, occurs in both free form and as an ester linked to lignin in seeds, leaves, and cell wall of plants (Mathew and Abraham 2004). Ferulic acid is used in the production of biodegradable polymers, which has a great potential as an opening material for the synthesis of natural flavour and aroma compounds. Ferulic acid is a precursor of 4-vinylguaiacol, one of the important aromatic flavors used in the beverages, pharmaceutical, and food industries (Mathew and Abraham 2006). Volatile phenol, 4-vinylguaiacol is often identified in beer, wine, and whiskey upon fermentation and brewing processes.

The microbial ferulic acid decarboxylase, which decarboxylate ferulic acid with simultaneous production of 4-vinylguaiacol, is supposed to be responsible for the detoxification of ferulic acid (Cavin et al. 1997, 1998; Clausen et al. 1994; Coghe et al. 2004; Degrassi et al. 1995; Huang et al. 1994; Smit et al. 2003). Mostly, phenolic acids that retard the growth of microorganisms (Huang et al. 2012a, b) are proposed to produce ferulic acid decarboxylase in response to 4-hydroxycinnamic acids.

The major metabolic step of ferulic acid bioconversion by crude cell extract of *Bacillus cereus* SAS-3006 was described in previous study (Mishra et al. 2014b). In this study, ferulic acid decarboxylase which plays a key role in the degradation of ferulic acid, was purified and partially characterized.

Materials and Methods

Ferulic acid (99%) and 4-vinylguaiacol (99%) used in this experiment were purchased from Sigma-Aldrich (GmbH, UK). High-performance liquid chromatography (HPLC) grade methanol and formic acid were purchased from Hi-Media (India). Other chemicals were of analytical grade.

Microorganism and Culture Conditions

The source of enzyme was *Bacillus cereus* SAS-3006. The minimal medium was constructed in the same approach as described by Muheim and Lerch (1999). The pH was adjusted to 7.0. All the carbon sources were filter sterilized by a 0.20 mm nylon filter (Sartorius, Minisart) before their addition to medium.

Cell Extract Preparation

The crude cell extract preparation method was same as describe by Mishra et al. (2014b).

Assay of Protein Concentrations

Protein concentrations were estimated by Bradford method (Bradford 1976), with bovine serum albumin as the standard.

Enzyme Assay

Ferulic acid decarboxylase activity was assayed by measuring the formation of the enzymatic product 4-vinylguaiacol (4-hydroxy-3-methoxystyrene) by HPLC and spectrophotometer method as describe previously (Mishra et al. 2014b).

Protein Purification

The enzyme purification steps were performed at 4 °C. The cell extract from 600 mL bacterial culture was dialyzed in 5 mL of 40 mM tris-HCl buffer by ultrafiltration through a 15 kDa cutoff membrane (Amicon Ultra-15 Centrifugal Filter, Millipore, USA) at pH 7.6. The dialyzed sample was loaded onto a column of DEAE-Cellulose equilibrated with 50 mM phosphate buffer at 7.3 pH. The enzyme was eluted by a 100 mL linear gradient from 0.0 to 0.5 M KCl in 50 mM phosphate buffer at 7.3 pH (Huang et al. 1994). Effluents of column were collected with fraction sizes of 2 mL at a flow rate of 1 mL/min. All fractions were monitored at 260 nm using spectrophotometer (Shimadzu, UV-1800). From elution pattern, the absorption peaks were pooled and concentrated to a small volume using Amicon membrane. The decarboxylase activity of partially purified cell extract to transform ferulic acid into 4-vinylguaiacol was studied at 37 °C. The total reaction mixture (1 mL) contained 50 mM sodium phosphate buffer (pH 6.0), ferulic acid (5.0 μg), and cell extract (50 μL). The reaction mixture was incubated for 0, 1 and 3 h. After incubation, the reaction was stopped by adding equal amount of acetic acid: methanol (1:4) to the reaction mixture. The samples were then analyzed by spectrophotometer and high-performance liquid chromatography for the detection of enzymatic products.

Characterization of Purified Enzyme

Effect of different concentration of substrate

To determine the stability of purified enzyme, different concentration of substrate (1.0, 2.5, 5.0, 7.5, 10.0, and 12.5 μg/mL) were incubated in the 50 mM sodium phosphate buffer (pH 6.0) at 37 °C for 0 and 1 h (Degrassi et al. 1995).

Determination of PH and Temperature Optima for Enzyme Activity

To check the pH stability, purified enzyme was incubated in sodium phosphate buffer (pH 5.0, 6.0 and 7.0) 37 °C for 0 and 1 h. To determine the thermal stability of purified enzyme, different temperatures (28, 37, and 45 °C) were used at pH 6.0 for 0 and 1 h. The remaining decarboxylase activity was assayed under standard conditions described above (Degrassi et al. 1995).

Effect of Metals Ions on Enzyme Activity

The effects of various metallic ions such as Fe^{2+}, Fe^{3+}, Zn^{2+}, Ca^{2+}, Mg^{2+}, and Mn^{2+} were applied in the reaction mixture at a concentration of 2.0 mM and activity was measured as described in the enzyme assay part (Degrassi et al. 1995).

Results and Discussion

Protein Purification

Ferulic acid decarboxylase appears to be inherent in the cell extracts when *Bacillus cereus* SAS-3006 cultivated in medium which is supplemented with ferulic acid. Maximum enzyme activity was observed in cultures at 3 h of growth. Cells grown were harvested by centrifugation, disrupted by ultrasonication, and subjected to the sequence of purification steps that are summarized in Table 1. Total yield of 22% of the activity initially found in cell extracts and an enzyme preparation purified 72.4 fold.

Table 1 Purification of ferulic acid decarboxylase from *Bacillus cereus* SAS-3006

Partial purification step	Total protein (mg)	Total activity (U)	Specific activity (U/mg)	Purification (fold)	Yield (%)
Crude extract	495	47.6	0.096	1	100
DEAE-cellulose	74	26.2	0.354	3.7	55
Ultrafilteration	1.51	10.5	6.95	72.4	22

Decarboxylase Production and Activity

Bacillus cereus SAS-3006 produces ferulate decarboxylase under nutrient limited conditions. This enzyme transforms ferulic acid to 4-vinylguaiacol as demonstrated by UV spectroscopy (Fig. 1), via decarboxylation. Additionally, HPLC analysis showed similar retention times for 4-vinylguaiacol and the degradation product of ferulic acid in *Bacillus cereus* SAS-3006 culture (data not shown). The reaction is catalyzed by a stable ferulic decarboxylase present in the crude cell extract from an induced *Bacillus cereus* SAS-3006 culture. The production of this enzyme seems to be induced by the presence of ferulic acid in the culture broth.

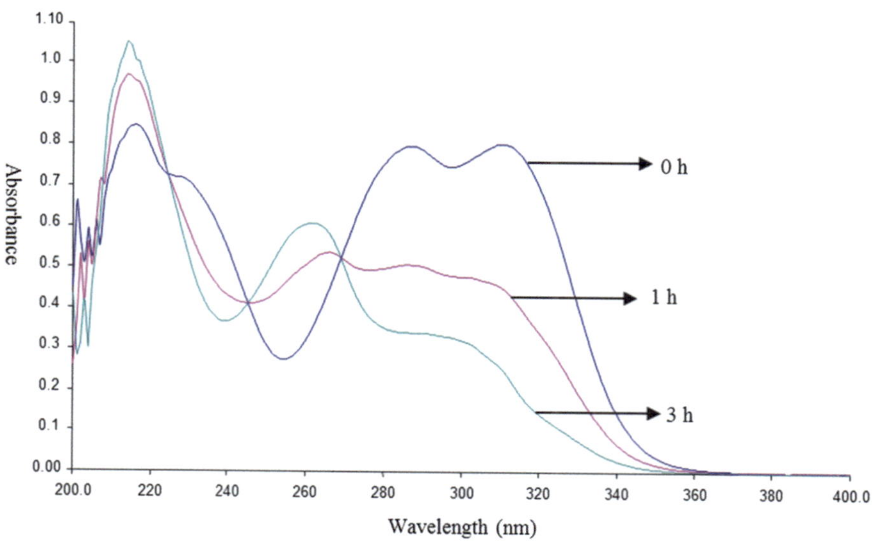

Fig. 1 Decrease of ferulic acid in the enzyme assay mixtures containing crude cell extract of *Bacillus cereus* SAS-3006 after 0–3 h of incubation

Characterization of Purified Enzyme

Effect of different concentration of substrate

When the purified enzyme (25 μL) was incubated with different concentration of substrate (1.0, 2.5, 5.0, 7.5, 10.0, and 12.5 μg/mL) in 50 mM sodium phosphate buffer at pH 6.0 and 37 °C temperature after 1 h of incubation. It was observed that, 2.5 μg/mL concentration of ferulic acid was found to be optimum for maximum 4-vinylguaiacol production. There was an increase in enzyme activity with increase in concentration of ferulic acid up to 2.5 μg/mL (Table 2). With further increase in concentration (5.0, 7.5, 10.0 and 12.5 μg/mL), there was a decrease in enzyme activity (Fig. 2). The enzyme activity is 0.333 U at 37 °C with 2.5 μg/mL concentration of ferulic acid.

Effect of pH and Temperature

Purified enzyme (25 μL) was incubated with ferulic acid (2.5 μg/mL) in 50 mM sodium phosphate buffer at different temperature 28, 37, and 45 °C for 1 h (Fig. 3). It was observed that, 37 °C temperature was found to be optimum for maximum 4-vinylguaiacol production.

When purified enzyme (25 μL) was incubated with 2.5 μg/mL of substrate at different temperature (28, 37, and 45 °C) in 50 mM sodium phosphate buffer at pH 6.0 after 1 h of incubation. It was observed that, 37 °C temperature was found to be optimum for maximum 4-vinylguaiacol production. The enzyme activity was high at 37 °C (Table 3).

Enzyme activity is optimal at 37 °C and 2.5 μg/mL concentration of ferulic acid as substrate. Kinetic studies indicated a K_m of 0.01 mM and $V_{\max}$ of 0.309 for 4-vinylguaiacol (Fig. 4).

Purified enzyme (25 μL) was incubated with 2.5 μg/mL ferulic acid in 50 mM sodium phosphate buffer at pH 5.0, 5.5, 6.0, 6.5, 7.0, 7.5, 8.0, and 8.5 at 37 °C for 1 h. It was observed that 6.5 was found to be best for maximum 4-vinylguaiacol production from the range of pH 6.5 to 7.5 (Fig. 5).

Table 2 Activity of purified enzyme when incubated with different concentrations of substrate

Conc. of substrate (μg/mL)	Enzyme activity (U)
1.0	0.0864
2.5	0.3339
5.0	0.3039
7.5	0.2099

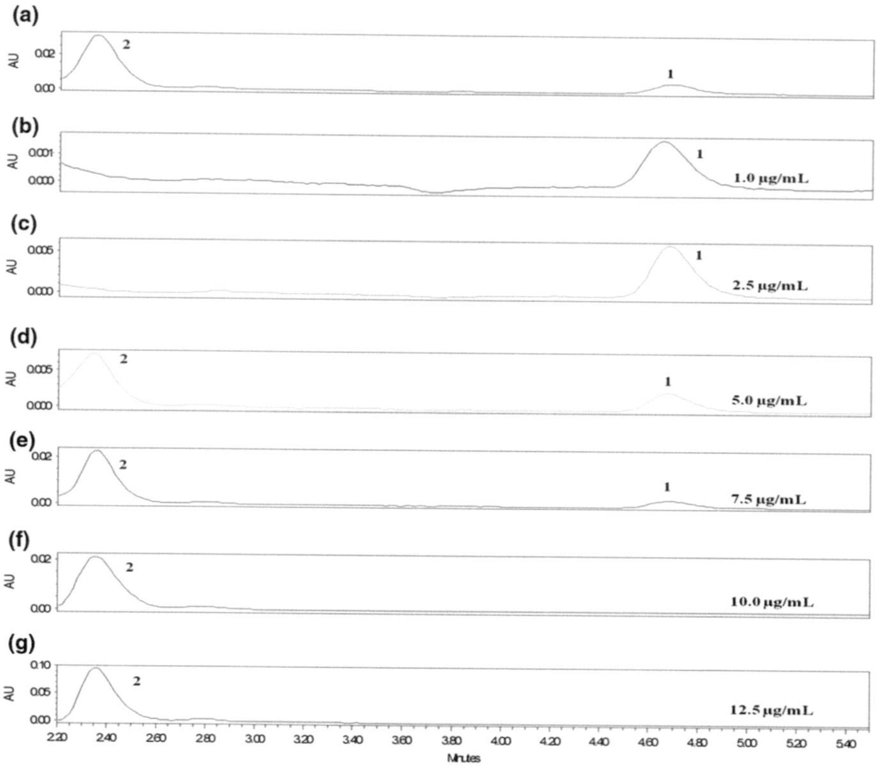

Fig. 2 Stack plot of HPLC chromatogram at 254 nm showing the in-vitro conversion of ferulic acid to 4-vinylguaiacol by purified enzyme of SAS-3006. Chromatogram (**a**) represents ferulic acid (*2*) and 4-vinylguaiacol (*1*) as standards. Chromatogram (**b**) and (**c**) represents decrease in the amount of ferulic acid with increase amount of 4-vinylguaiacol when supplemented with the 1.0 and 2.5 μg/mL ferulic acid respectively. Chromatogram (**d**), (**e**), (**f**) and (**g**) represents increase in concentration of ferulic acid (5.0, 7.5, 10.0 and 12.5 μg/mL), with decrease in the product formation on 1st hour of incubation respectively

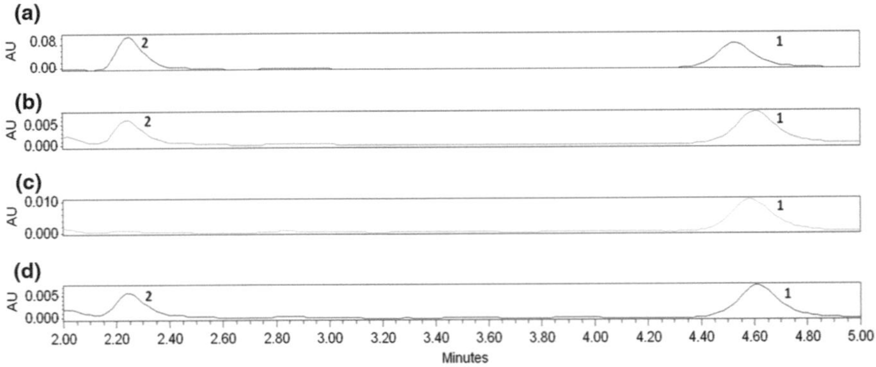

Fig. 3 Stack plot of HPLC chromatogram at 254 nm showing the in-vitro conversion of ferulic acid to 4-vinylguaiacol by purified enzyme of SAS-3006. Chromatogram (**a**) represents ferulic acid (*2*) and 4-vinylguaiacol (*1*) as standards. Chromatogram (**b**), (**c**) and (**d**) represents 4-vinylguaiacol formation when supplemented with the same at 28, 37 and 45 °C on 1st hour of incubation respectively

Table 3 Activity of purified enzyme when incubated with 2.5 μg/mL at different temperature

Incubation temp (°C)	Enzyme activity (U)
28	0.298
37	0.309
45	0.278

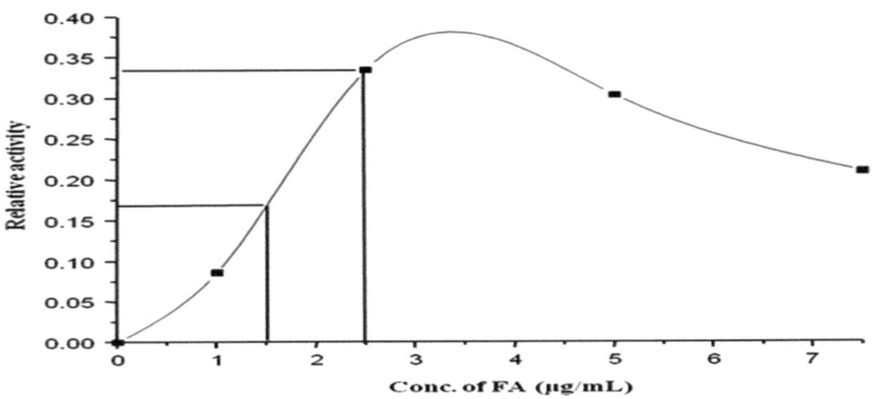

Fig. 4 Kinetic studies indicated a K_m of 0.01 mM and V_{max} of 0.309 for 4-vinylguaiacol

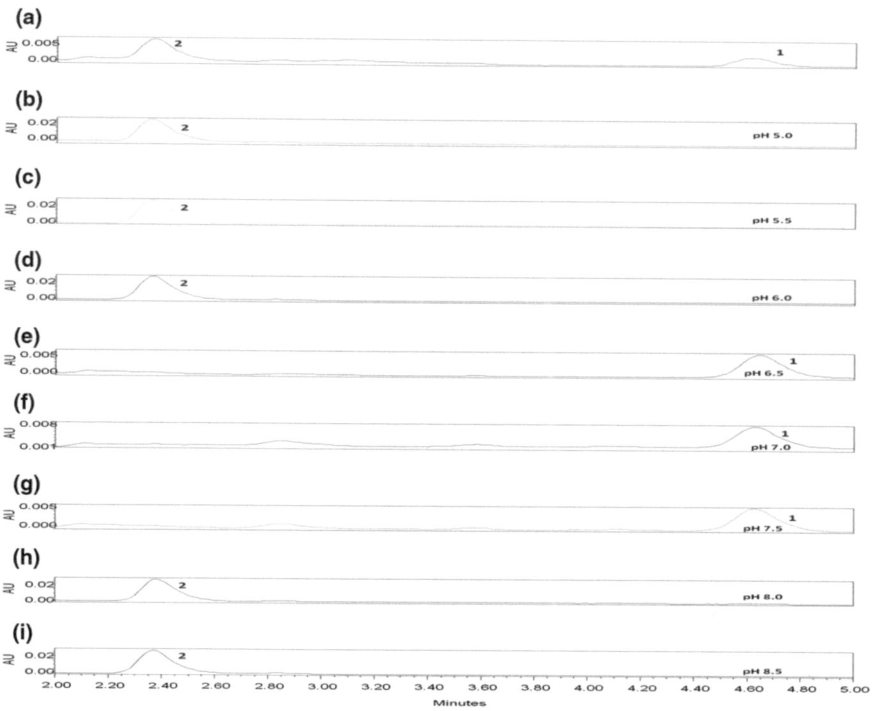

Fig. 5 Stack plot of HPLC chromatogram at 254 nm showing the in-vitro conversion of ferulic acid to 4-vinylguaiacol by purified enzyme of SAS-3006. Chromatogram (**a**) represents ferulic acid (*2*) and 4-vinylguaiacol (*1*) as standards. Chromatogram (**b**), (**c**), (**d**), (**e**), (**f**), (**g**), (**h**) and (I) represents 4-vinylguaiacol formation when supplemented with the same at pH range from 5.0 to 8.5 on 1st hour of incubation respectively

Effect of Metals Ions on Enzyme Activity

When purified enzyme (25 μL) was incubated with 2.5 μg/mL of ferulic acid at pH 6.5 and 37 °C in 50 mM sodium phosphate buffer by addition of metals ions (Fe^{2+}, Fe^{3+}, Ca^{2+}, Mg^{2+}, Mn^{2+}, and Zn^{2+} at 2.0 mM concentration) on first h of incubation. It was observed that, enzyme activity was not affected by metals ions (Fig. 6).

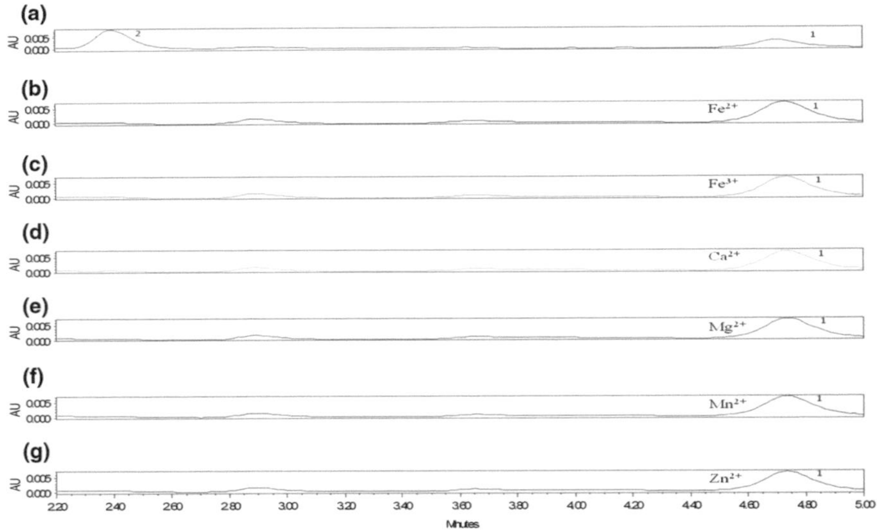

Fig. 6 Stack plot of HPLC chromatogram at 254 nm showing the in-vitro conversion of ferulic acid to 4-vinylguaiacol by purified enzyme of SAS-3006. Chromatogram (**a**) represents ferulic acid (*2*) and 4-vinylguaiacol (*1*) as standards. Chromatogram (**b**), (**c**), (**d**), (**e**), (**f**) and (**g**) represents no effect on enzyme activity with addition of metals ions such as Fe^{2+}, Fe^{3+}, Ca^{2+}, Mg^{2+}, Mn^{2+} and Zn^{2+} at 2.0 mM concentration on 1st hour of incubation respectively

Conclusion

This is the first report of the purification of a ferulate decarboxylase from a *Bacillus cereus* SAS-3006. Ferulic acid decarboxylase is efficiently capable to convert ferilic acid to 4-vinylguaiacol. The purification and characterization of ferulic acid decarboxylase provide a better understanding of the requirements for catalysis. Decarboxylation represents the first step of a degradative pathway, like the one suggested for *Fusarium solani* by Nazareth and Mavinkurve (1986).

Acknowledgements This work is financially supported by University Grants Commission for the Major Research Project [F.No. 37-115/2009 (SR)]. Shashank Mishra gratefully acknowledges the Department of Bio-Engineering, Birla Institute of Technology, Mesra, Ranchi, for providing the infrastructure facilities. The authors acknowledge Dr. Sanjaya Swain, Research Officer, Central Instrumentation Facility (CIF), BIT, Mesra for providing analytical facilities.
Declaration of interest: The authors report no declarations of interest. The authors alone are responsible or the content and writing of the paper.

References

Bradford MM (1976) A rapid and sensitive method for the quantitation of microgram quantities of protein utilizing the principle of protein-dye binding. Anal Biochem 72:248–254

Cavin JF, Barthelmebs L, Guzzo J, Beeumen JV, Samyn B, Travers JF, Diviès C (1997) Purification and characterization of an inducible p-coumaric acid decarboxylase from *Lactobacillus plantarum*. FEMS Microbiol Lett 147:291–295

Cavin JF, Dartois V, Divie`s C (1998) Gene cloning, transcriptional analysis, purification, and characterization of phenolic acid decarboxylase from *Bacillus subtilis*. Appl Environ Microbiol 64:1466–1471

Clausen M, Lamb CJ, Megnet R, Doerner PW (1994) PAD1 encodes phenylacrylic acid decarboxylase which confers resistance to cinnamic acid in *Saccharomyces cerevisiae*. Gene 142:107–112

Coghe S, Benoot K, Delvaux F, Vanderhaegen B, Delvaux FR (2004) Ferulic acid release and 4-vinylguaiacol formation during brewing and fermentation: indications for feruloyl esterase activity in *Saccharomyces cerevisiae*. J Agric Food Chem 52:602–608

Degrassi G, Polverino de Laureto P, Bruschi CV (1995) Purification and characterization of ferulate and p-coumarate decarboxylase from *Bacillus pumilus*. Appl Environ Microbiol 61:326–332

Giedraityte G, Kalediene L (2014) Biotransformation of eugenol via protocatechuic acid by thermophilic *Geobacillus* sp. AY 946034 strain. J Microbiol Biotechnol 24:475–482

Huang Z, Dostal L, Rosazza JPN (1994) Purification and characterization of a ferulic acid decarboxylase from *Pseudomonas fluorescens*. J Bacteriol 176:5912–5918

Huang HK, Tokashiki M, Maeno S, Onaga S, Taira T, Ito S (2012a) Purification and properties of phenolic acid decarboxylase from *Candida guilliermondii*. J Ind Microbiol Biotechnol 39:55–62

Huang HK, Chen LF, Tokashiki M, Ozawa T, Taira T, Ito S (2012b) An endogenous factor enhances ferulic acid decarboxylation catalyzed by phenolic acid decarboxylase from *Candida guilliermondii*. AMB Express 2:1–10

Mathew S, Abraham TE (2004) Ferulic acid: an antioxidant found naturally in plant cell walls and feruloyl esterases involved in its release and their applications. Crit Rev Biotechnol 24:59–83

Mathew S, Abraham TE (2006) Bioconversions of ferulic acid, a hydroxycinnamic acid. Crit Rev Microbiol 32:115–125

Mishra S, Sachan A, Vidyarthi AS, Ghosh Sachan S (2014a) Microbial production of 4-vinylguaiacol from ferulic acid by *Bacillus cereus* SAS-3006. Biocatal Biotransform 32:259–266

Mishra S, Sachan A, Vidyarthi AS, Ghosh Sachan S (2014b) Transformation of ferulic acid to 4-vinyl guaiacol as a major metabolite: a microbial approach. Rev Environ Sci Biotechnol 13:377–385

Muheim A, Lerch K (1999) Towards a high-yield bioconversion of ferulic acid to vanillin. Appl Microbiol Biotechnol 51:456–461

Nazareth S, Mavinkurve S (1986) Degradation of ferulic acid via 4-vinylguaiacol by *Fusarium solani*. Can J Microbiol 32:494–497

Smit A, Cordero Otero RR, Lambrechts MG, Pretorius IS, van Rensburg P (2003) Enhancing volatile phenol concentrations in wine by expressing various phenolic acid decarboxylase genes in *Saccharomyces cerevisiae*. J Agric Food Chem 51:4909–4915

Marker-Assisted Breeding of Recombinant 1RS.1BL Chromosome for Improvement of Bread Making Quality and Yield of Wheat (*Triticum aestivum* L.)

Rajdeep Kaur, Pritesh Vyas, Prachi Sharma, Imran Sheikh, Rahul Kumar and H.S. Dhaliwal

Abstract

The 1RS.1BL translocation has been extensively used as a source of genes on 1RS for multiple disease resistance (*Lr26*, *Yr9*, *Sr31* and *Pm8*) and enhanced yield of wheat cultivars. However, cultivars with the 1RS.1BL translocation have sticky dough due to the presence of *Sec-1* on 1RS and absence of *Glu-B3/Gli-B1* on 1BS. Many QTL useful for root traits have also been mapped on 1RS. A tertiary recombinant 1RS line lacking *Sec-1* but with useful root QTL and *Glu-B3/Gli-B1* of 1BS has been developed at the University of California, Riverside USA which is being used for marker-assisted breeding of wheat cultivars for higher yield and improved bread making quality. PCR-based SSR markers (ω-sec-P3/P4, Rye R3/F3, Psp3000, Sfr 43) were used to check the presence of *Sec-1*, rye translocation, *Glu-B3/Gli-B1* and *Pm8* for characterization of wheat cultivars and 1RS recombinants, respectively. The markers amplified the desired amplicons and were well synchronized to each other about the presence and absence of various genes in each cultivar and recombinants. The absence of *Pm8* and poor root traits in MA1Pavon confirmed the close association between *Pm8* and useful root QTL. Two of the recombinants, i.e., 1B+38 and 1RS44:38 were found to have *Pm8* gene as well as better root traits than in MA1Pavon. The SDS micro-sedimentation test of various 1RS recombinants and parental lines showed that 1RS–1BL with *Sec-1* had lowest micro-sedimentation test (MST) value while addition of *Glu-B3/Gli-B1* had negligible improvement. These findings suggest that the MST values are not much affected by the absence of *Glu-B3/Gli-B1* and hence the lines having better root traits with no *Glu-B3/Gli-B1* and secalin could be used for improvement of bread making quality and yield in wheat.

R. Kaur · P. Vyas · P. Sharma · I. Sheikh · R. Kumar · H.S. Dhaliwal (✉)
Akal College of Agriculture, Eternal University, Baru Sahib, Himachal Pradesh, India
e-mail: hsdhaliwal07@gmail.com

K. Mukhopadhyay et al. (eds.), *Applications of Biotechnology for Sustainable Development*, DOI 10.1007/978-981-10-5538-6_21

Keywords
Triticum aestivum · 1RS.1BL · Root traits · Micro-sedimentation · *Pm8* · MA1Pavon · Marker-assisted breeding

Introduction

Wheat (*Triticum aestivum* L.) is an important staple food estimated to account for approximately 20% of all human food calories consumed worldwide (FAOSTAT 2014). It is produced all over the world and estimated production in 2014 was 731.2 million tons. Bread wheat is one of the world's three main cereal crops, along with rice and maize, with best grains for bread making. The introgression of alien chromatin from rye (*Secale cereale* L.) into the genetic background of wheat has been exploited in wheat breeding programs. The 1RS.1BL translocation, which originated from a spontaneous translocation (Zeller 1973) of 1RS of rye, has been useful source of genes for disease resistance (*Lr26*, *Yr9*, *Sr31* and *Pm8*) and enhanced yield in wheat (Villareal et al. 1998; Waines and Ehdaie 2007). However, the dough prepared from 1RS.1BL lines is unacceptable for bread making because of its excessive stickiness and mixing intolerance (Dhaliwal et al. 1990; Lee et al. 1995) due to the secalin proteins (Graybosch et al. 1993).

Several attempts have been made to improve quality of the 1RS translocations in wheat (Koebner et al. 1986; Millet and Feldman 1993; Lukaszewski 1993, 1997). Centric translocation of short arm of rye chromosome was induced by *ph1b* mutation to recombine with the short arms of wheat group-1 chromosomes. The *Glu-B3*/*Gli-B1* loci of wheat are non-homoeoallelic to the *Sec-1* locus of rye and are separated by about a 13 cM long segment, containing disease resistance loci (Lukaszewski 2000).

The chromosomal manipulations were performed in *T. aestivum* cv. Pavon 76 amber wheat from the International Wheat and Maize Improvement Center (CIMMYT) Mexico now referred to as Pavon. After a series of manipulations and development of recombinants a tertiary recombinant MA1Pavon was developed where the *Sec-1* locus was removed and *Glu-B3*/*Gli-B1* were retained (Lukaszewski, 2000).

Roots, the hidden half of a plant, are important for numerous functions including water and nutrient uptake that make it difficult to overlook their importance to plant productivity (MacMillan et al. 2006; Bishopp and Lynch 2015). A total of 15 QTL effects (6 additive and 9 epistatic) were detected for different traits of root length and root weight in 1RS wheat. Some of these QTLs are tightly linked to marker *Pm8*, a powdery mildew resistance locus. These were detected between two adjacent regions marked on the map by loci *Pm8* and *Glu-B3*/*Gli-B1* (Sharma et al. 2011).

The results of an integrated study including molecular marker validation, study of root phenotype, and analysis of gluten strength reported in this article have led to selection of a suitable 1RS.1BL recombinant having all the desired root traits and bread making quality required to be introgressed in commercial wheat cultivars through marker-assisted breeding.

Materials and Methods

DNA Isolation and Validation of Markers

DNA was isolated and quantified from the leaf tissue of 19 wheat genotypes using CTAB method (Murray and Thompson 1980) with minor modifications. The presence of ω-secalins, *Glu-B3/Gli-B1*, 1RS translocation and *Pm8* genes in different wheat genotypes and genetic stocks by several PCR-based markers was validated (Table 1).

Phenotyping of Roots

Root phenotyping was performed using different root traits including length, branching, and biomass from the growing seedlings of the genotypes. The experimental system included the selected wheat cultivars and stocks which were grown under two conditions. In one, acid washed sand was used with the half strength of Hoagland medium (Hoagland and Arnon 1950) with some modifications. In the other condition, soil and vermicompost mix in 50:50 proportion of weight were used. Seeds were sown in transparent pots which were covered with black polythene to avoid light interference. Transparent pots enabled us to monitor the root growth, root length, and branching pattern among the various cultivars and recombinants as and when required. Root traits were recorded after 15 days of germination.

Table 1 PCR primers used for amplification of different genes/regions

Primers	Forward and reverse sequence	Amplicon length	Annealing temperature (°C)	Reference
Rye R3/F3 (rye translocation)	F: GATCGCCTCTTTTGCCAAGA R: TCACTGATCACAAGAGCTTG	1.4 kb	57	Katto et al. (2004)
Omega P3–P4 (secalins)	F:GTTTGCTGGGGAATTATTTG R:TCCTCATCTTTGTCCTCGCC	412 bp	66	Froidmont (1998)
Psp3000 (*GluB3/Gli B1*)	F: GCA GACCTG TGTCATTGG TC′ R:GATATAGTGGCAGCAGGATACG	300–500 bp	64	Devos et al. (1995)
sfr43(*Pm8*)	F: TGGCTTCCAACAGCCCTAGC R:AGGCTTTTGCACCTTCTCT	662 bp	64	Hurni et al (2013)

Micro-Sedimentation Test

Micro-sedimentation test was performed from the matured seeds of different 1RS.1BL translocated and non-translocated lines with presence and absence of secalin and *Glu-B3/Gli-B1* grown in field and greenhouse conditions. The experiment was performed according to the method of Dick and Quick (1983) with some modifications.

Results and Discussion

In this study, different wheat cultivars and recombinants with or without 1RS.1BL translocation were used. PCR markers linked to different genes/regions of 1RS.1BL translocation were optimized for amplification studies. Micro-SDS sedimentation test values and root traits for all the recombinants and cultivars were examined using various parameters.

Molecular Maker Analysis

The primer pair used to identify the introgression of rye segment in the parental and recombinant lines of wheat amplified a band of 1.4 kb bp (Katto et al. 2004) and showed the presence of rye segment in MA1Pavon, 1B+38, 1RS44:38, 1RS40:9, UP2338, HS240, PBW 343GpcB1+*Lr24* and there was no amplification in those without 1RS.1BL translocation that are Pavon, PBW 343 with *Lr24*, *Lr28* and *Yr15* (Fig. 1a). Primer ω-sec P3/P4 was used to identify the presence of ω-secalins in the parental and recombinant lines of wheat which amplified a fragment of 412 bp (Froidmont 1998; Chai et al. 2006) showing the presence of ω-secalins in 1RS 40:9, UP2338, HS240, PBW 343GpcB1+*Lr24* while no desired amplification was

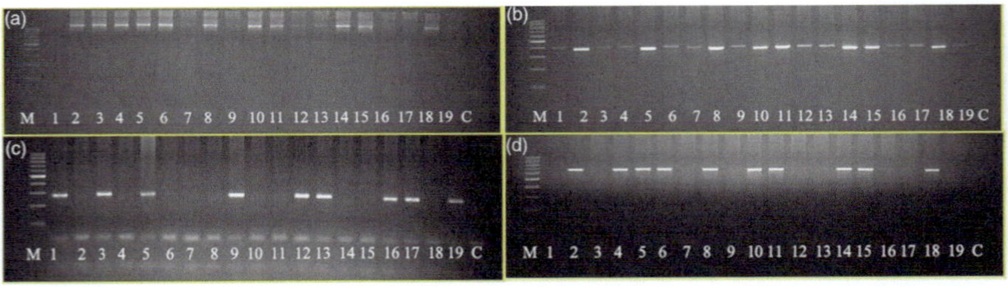

Fig. 1 Agarose gel profiling of PCR-amplified products using primer pair for **a** Rye Translocation **b** ω-secalins **c** *Glu-B3/Gli-B1* **d** *Pm8* M. Mol. wt Marker, *1* Pavon, *2* Pavon 1RS–1BL, *3* MA1Pavon, *4* 1B+38, *5* 1RS40:9, *6* 1RS 44:38, *7* Pavon DtBL, *8* UP2338, *9* HD2967, *10* HS240, *11* PBW 343GpcB1+*Lr24*, *12* PBW 343 with *Lr24*, *Lr28* and *Yr15*, *13* Pavon 1-96-1, *14* MAC2496, *15* MAC6222, *16* MAC6478, *17* HS507, *18* WMB3617, *19* WMB3618, *C* Control

observed in Pavon, MA1Pavon, 1B+38, 1RS44:38, PBW 343 with *Lr24*, *Lr28* and *Yr15* (Fig. 1b) indicating the absence of *Sec-1* genes in these lines.

The Primer Psp3000 was used to identify the presence of *Glu-B3/Gli-B1* in the parental and recombinant lines of wheat. It amplified a band of 300 bp (Devos et al. 1995; Howell et al. 2014) showing the presence of *Glu-B3/Gli-B1* in 1RS40:9, Pavon, MA1Pavon, PBW 343 with *Lr24*, *Lr28* and *Yr15* while no amplification was observed in 1B+38, 1RS44:38, UP2338, HS240, PBW 343GpcB1+*Lr24* (Fig. 1c) indicating the absence of *Glu-B3/Gli-B1* gene in these lines. Primer Sfr 43 was used to identify the presence of *Pm8* in the parental and recombinant lines of wheat with amplification fragment of 662 kb (Hurni et al. 2013) showing the presence of *Pm8* as well as linked root QTLs in 1RS40:9, 1B+38, 1RS44:38, UP2338, HS240, PBW 343GpcB1+*Lr24*. There was no amplification in Pavon, MA1Pavon, PBW 343 with *Lr24*, *Lr28* and *Yr15* (Fig. 1d). Interestingly *Pm8* was found absent in MA1Pavon indicating that it was within the introgressed wheat chromatin controlling *Glu-B3/Gli-B1*. The short arm (1RS) of rye on the distal end has three tightly linked genes, i.e., *Lr26* (leaf rust resistance), *Yr9* (stripe rust resistance) and *Sr31* (stem rust resistance) which segregate away from *Pm8* (Liu et al. 2015). *Glu-B3/Gli-B1* is present in MA1Pavon so it is evident from the present study that introgression of *Glu-B3/Gli-B1* may be the reason for the absence of *Pm8* gene during introgression in this line.

Micro-Sedimentation Test

Under field conditions, high MST values were obtained for MA1Pavon, Pavon, 1RS44:38, and 1B+38 while very low values were recorded for soft wheat cultivars HPW89 and Naphal (Fig. 2). Moderately high values were found in UP2338, HD2967, and PBW 343 with *Lr24*, *Lr28* and *Yr15*. Under greenhouse conditions, high MST values were obtained for Pavon 1-96-1, Pavon DtBL, and Pavon 76 without 1RS and secalin. Moderately high values were obtained for Pavon 1RS.1BL and Pavon 1RS44:38.1BL (Fig. 3).

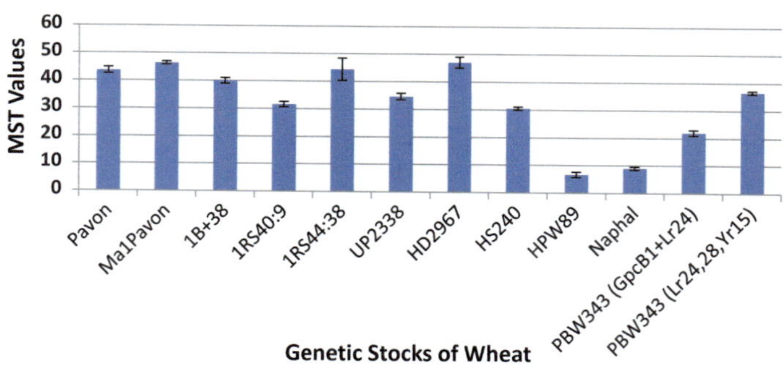

Fig. 2 Graphical representation of MST values of different cultivars and genetic stocks of wheat grown under field conditions

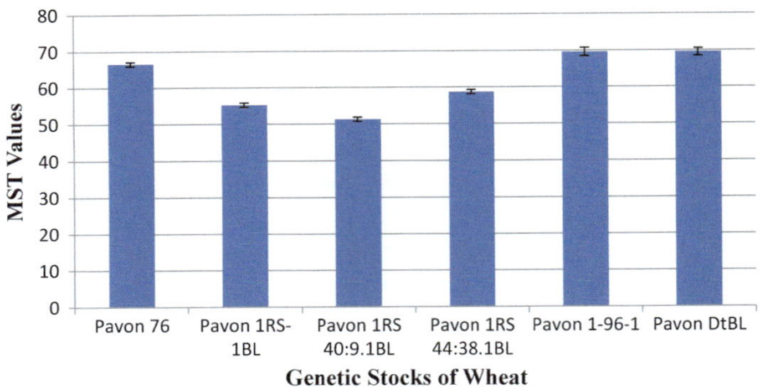

Fig. 3 Graphical representation of MST values of different genetic stocks of wheat grown in greenhouse

The presence of 1RS with *Sec-1* in the wheat cultivars and lines had poor MST values which were only marginally improved by retaining wheat *Glu-B3/Gli-B1* indicating the secalin had major adverse effect on bread making characteristics and not the absence of *Glu-B3/Gli-B1*. Hence only the recombinants without *Glu-B3/Gli-B1* such as 1RS44:38 could be more useful for improving root traits than MA1Pavon. The presence of secalins adversely affects the bread making quality by giving sticky and hard dough (Barbeau et al. 2003). However, Burnett et al. (1995) did not find any significant difference in the dough quality of 1B/1R and their 1B near isolines although the sedimentation values were low for 1B/1R lines as observed here.

Comparison of Root Traits

The agronomic traits comprising of shoot biomass, root biomass, root length, and root branching patterns were tested under different conditions which revealed some valuable information (Tables 2 and 3). The Hoagland medium had best expression of agronomic root traits in comparison to vermicompost soil mix (Tables 2 and 3). The dry weight of roots was found to be of higher magnitude in all lines in Hoagland medium in comparison to vermicompost soil mix. The highest root dry weight was in 1B+38 line 26.10 mg/plant/seedling and lowest in 1RS40:9 (11.7 mg/plant/seedling). In case of vermicompost soil mix the highest dry matter of root was recorded in 1RS44:38 of 5.92 mg/plant/seedling and the lowest in HS240 of 3.06 mg/plant/seedling (Tables 2 and 3). The wet weight of seedlings showed that the Hoagland medium conditions suited better to all the lines than the vermicomposting conditions. The wet weight of seedlings was found to be the best in Pavon with a value of 538.8 mg/plant/seedling while 1RS 44:38 had the highest value of 329.4 mg/plant/seedling in vermicomposting conditions (Tables 2 and 3). No significant difference was observed in branching patterns in Hoagland medium

Table 2 Different root parameters of various genetic stocks of wheat during their growth in Hoagland medium

Genetic stock	Length (cm)	Branching	Weight of seedling (mg)	Wet weight roots (mg)	Dry weight of roots (mg)
Pavon	14.2 ± 2.30	3.8 ± 0.83	538.8 ± 1.45	99.7 ± 3.33	12.6 ± 1.24
MA1 Pavon	14.5 ± 2.10	4.0 ± 0.54	437.5 ± 5.74	101.6 ± 3.39	11.1 ± 1.49
1B+38	19.8 ± 2.57	4.4 ± 0.54	501.4 ± 0.84	205.8 ± 6.79	26.10 ± 1.50
1RS40:9	13.0 ± 1.76	3.8 ± 0.44	280.4 ± 1.68	76.7 ± 1.36	11.7 ± 0.34
1RS44:38	18.9 ± 2.94	4.4 ± 0.54	494.4 ± 6.48	131.5 ± 4.52	22.2 ± 2.65
UP2338	17.3 ± 3.30	3.2 ± 0.44	429.6 ± 3.12	109.0 ± 3.91	19.3 ± 1.81
HD2967	10.0 ± 1.32	3.0 ± 0.23	354.5 ± 2.10	152.5 ± 2.32	15.2 ± 1.78
HS240	12.2 ± 2.31	3.0 ± 1.22	276.9 ± 1.52	59.7 ± 0.23	20.2 ± 2.13

Each reading is the mean of five replicates with standard deviation

Table 3 Different root parameters of various genetic stocks of wheat during their growth in vermicompost soil mix

Genetic stock	Length (cm)	Branching	Weight of seedling (mg)	Wet weight of roots (mg)	Dry weight of roots (mg)
Pavon	11.1 ± 2.53	3.4 ± 0.54	209.7 ± 5.38	26.1 ± 4.22	4.80 ± 1.93
MA1 Pavon	13.7 ± 1.86	3.4 ± 0.54	267.4 ± 4.88	27.8 ± 3.38	5.12 ± 1.38
1B+38	12.6 ± 1.46	3.2 ± 0.44	282.9 ± 5.18	21.3 ± 4.91	5.22 ± 0.70
1RS 40:9	14.4 ± 3.25	4.0 ± 0.66	324.6 ± 1.84	31.1 ± 4.63	4.78 ± 0.69
1RS 44:38	13.9 ± 3.38	4.6 ± 0.89	329.4 ± 2.75	42.2 ± 1.60	5.92 ± 1.45
UP 2338	18.5 ± 3.23	3.2 ± 0.44	280.0 ± 1.05	42.2 ± 1.30	4.88 ± 1.25
HD 2967	9.8 ± 2.05	3.0 ± 0.56	190.4 ± 4.71	23.0 ± 3.48	3.90 ± 1.13
HS 240	11.4 ± 2.49	3.0 ± 0.45	202.8 ± 5.67	20.4 ± 2.15	3.06 ± 0.60

Each reading is the mean of five replicates with standard deviation

and vermicomposting soil conditions. High values for root length were observed in Hoagland medium conditions. With an exception of 1RS40:9 and UP2338, all other lines showed better root length in Hoagland medium than under vermicomposting. Wheat cultivars with 1RS and recombinant 1RS.1BL chromosomes with and without *Sec-1* had better root growth parameters indicating that the QTL for superior root growth on 1RS are not associated with *Sec-1 locus* and hence recombinants without *Sec-1* can be used for marker-assisted breeding of 1RS.1BL. The usefulness of 1RS.1BL on wheat agronomic traits has been reported (Rajaram et al. 1983; Villareal et al. 1998; Moreno-Sevilla et al. 1995a, b; Monneveux et al. 2003). The 1RS arm had larger effect on root biomass (Howell et al. 2014; Sharma et al. 2009; Waines and Ehdaie 2007).

Furthermore, the gene *Pm8* for powdery mildew resistance is more likely linked to QTL for superior root traits than *Glu-B3/Gli-B1* and the genes *Lr26*, *Sr31* and *Yr9* for rust resistance (Sharma et al. 2011). One interesting fact that links up the two studies, i.e., the molecular marker study and the study of root traits is that the *Pm8* gene marker showed no amplification in MA1Pavon. *Pm8* gene possesses a tightly linked QTL for root traits such as total length of roots and number of roots (Sharma et al. 2011). These root traits provide strong agronomic benefits to the plant for overall development. Due to the absence of *Pm8* in MA1Pavon, the root traits were also affected. The protein content and the MST values for MA1Pavon with GluB3/GliB1 and 1RS44:38 without it did not show much difference. Recombinant line MA1 Pavon with wheat *Glu-B3/Gli-B1* had poor root parameters than 1RS44:38 and 1B+38 without *Glu-B3/Gli-B1* indicating that these are closely associated with 1RS of rye which are replaced by *Glu-B3/Gli-B1* of wheat and hence may not be useful for MAB of QTL for superior root traits. Higher yield and better canopy water status was reported in standard 1RS line and IRS^{RW} in which distal region was not replaced by that of wheat 1BS than the $1RS^{WW}$ NILs in which the distal and proximal part were replaced by that of wheat excluding secalin and including *Glu B3/GliB1* (Howell et al. 2014).

Conclusion

The two 1RS.1BL recombinants, i.e., 1B+38 and 1RS44:38 were found to have *Pm8* gene as well as better root traits than in MA1Pavon. The SDS micro-sedimentation test of various 1RS recombinants and parental lines showed that 1RS–1BL with *Sec-1* had lowest micro-sedimentation test (MST) value while substitution of *Glu-B3/Gli-B1* had negligible improvement. The present investigation suggests that MST values are not affected by the absence of *Glu-B3/Gli-B1*, and hence the lines having better root traits with no *Glu-B3/Gli-B1* and secalin could be used for improvement of bread making quality and yield in wheat.

Acknowledgements Authors acknowledges Department of Biotechnology (DBT), Govt. of India, New Delhi for providing research grant (BT/PR10886/AGII/106/934/2014). We also acknowledge Akal College of Agriculture for providing us the research facilities in the department.

References

Barbeau WE, Schwarzlaff SS, Uriyo MG, Johnson JM, Harris CH, Griffey CA (2003) Origin and practical significance of the sticky dough factor in 1BL/1RS wheats. J Sci Food Agri 83:29–38
Bishopp A, Lynch JP (2015) The hidden half of crop yields. Nat Plants 1:1–2
Burnett CJ, Lorenz KJ, Carver BF (1995) Effects of the 1B/1R translocation in wheat on composition and properties of grain and flour. Euphytica 86:159–166
Chai JF, Zhou RH, Jia JZ, Liu X (2006) Development and application of a new codominant PCR marker for detecting 1BL-1RS wheat-rye chromosome translocations. Plant Breed 125:302–304

Devos K, Bryan G, Collins A, Stephenson P, Gale M (1995) Application of two microsatellite sequences in wheat storage proteins as molecular markers. Theor Appl Genet 90:247–252

Dhaliwal A, Mares D, Marshall D (1990) Measurement of dough surface stickiness associated with the 1B/1R chromosome translocation in bread wheat. J Cereal Sci 12:165–175

Dick J, Quick J (1983) A modified screening test for rapid estimation of gluten strength in early-generation durum wheat breeding lines. Cereal Chem 60:315–318

FAOSTAT (2014) FAOSTAT. http://faostat.fao.org/in

Froidmont D (1998) A co-dominant marker for the 1BL/1RS wheat-rye translocation via multiplex PCR. J Cereal Sci 27:229–232

Graybosch RA, Peterson CJ, Hansen LE, Worrall D, Shelton D, Lukaszewski A (1993) Comparative flour quality and protein characteristics 1BL/1RS and 1AL/1RS wheat-rye translocation lines. J Cereal Sci 17:95–106

Hoagland DR, Arnon DI (1950) The water-culture method for growing plants without soil. Circ California Agric Exp Stn 347

Howell T, Hale I, Jankuloski L, Bonafede M, Gilbert M, Dubcovsky J (2014) Mapping a region within the 1RS.1BL translocation in common wheat affecting grain yield and canopy water status. Theor Appl Genet 127:2695–2709

Hurni S, Brunner S, Buchmann G, Herren G, Jordan T, Krukowski P, Wicker T, Yahiaoui N, Mago R, Keller B (2013) Rye *Pm8* and wheat *Pm3* are orthologous genes and show evolutionary conservation of resistance function against powdery mildew. Plant J 76:957–969

Katto MC, Endo TR, Nasuda S (2004) A PCR-based marker for targeting small rye segments in wheat background. Genes Genet Sys 79:245–250

Koebner RMD, Shepherd KW, Appels R (1986) Controlled introgression to wheat of genes from rye chromosome arm 1RS by induction of allosyndesis. 2 Characterisation of recombinants. Theor Appl Genet 73:209–217

Lee J, Graybosch R, Peterson C (1995) Quality and biochemical effects of a IBL/IRS wheat-rye translocation in wheat. Theor Appl Genet 90:105–112

Liu S, Rudd JC, Bai G, Haley SD, Ibrahim AM, Xue Q, Hays DB, Graybosch RA, Devkota RN, St Amand P (2015) Molecular markers linked to important genes in hard winter wheat. Crop Sci 54:1304–1321

Lukaszewski AJ (1993) Reconstruction in wheat of complete chromosomes 1B and 1R from the 1RS.1BL translocation of 'Kavkaz'origin. Genome 36:821–824

Lukaszewski AJ (1997) Further manipulation by centric misdivision of the 1RS.1BL translocation in wheat. Euphytica 94:257–261

Lukaszewski AJ (2000) Manipulation of the 1RS.1BL translocation in wheat by induced homoeologous recombination. Crop Sci 40:216–225

MacMillan K, Emrich K, Piepho HP, Mullins C, Price A (2006) Assessing the importance of genotype x environment interaction for root traits in rice using a mapping population. I: a soil-filled box screen. Theor Appl Genet 113:977–986

Millet E, Feldman M (1993) Deletion of the secalin gene *Sec-1* in 1BL/1RS line by gamma irradiation. In: Proceedings of the 8th international wheat genetics symposium, Beijing

Monneveux P, Reynolds M, Zaharieva M, Mujeeb-Kazi A (2003) Effect of T1BL. 1RS chromosome translocation on bread wheat grain yield and physiological related traits in a warm environment. Cereal Res. Comm. 31:371–378

Moreno-Sevilla B, Baenziger P, Peterson C, Graybosch R, McVey D (1995a) The 1BL/1RS translocation: agronomic performance of F_3-derived lines from a winter wheat cross. Crop Sci 35:1051–1055

Moreno-Sevilla B, Baenziger P, Shelton D, Graybosch R, Peterson C (1995b) Agronomic performance and end-use quality of 1B vs. 1BL/1RS genotypes derived from winter wheat 'Rawhide'. Crop Sci 35:1607–1612

Murray HG, Thompson WF (1980) Rapid isolation of high molecular weight DNA. Nucleic Acid Res 8:4321–4325

Rajaram S, Mann CE, Ortiz-Ferrara G, Mujeeb-Kazi A (1983) Adaptation, stability and high yield potential of certain 1B/1R CIMMYT wheats. In: Sadao Sakamoto (ed) Proceedings of the 6th international wheat genetics symposium. Plant Germplasm Institute, Faculty of Agriculture, Kyoto University, Kyoto

Sharma S, Bhat PR, Ehdaie B, Close TJ, Lukaszewski AJ, Waines JG (2009) Integrated genetic map and genetic analysis of a region associated with root traits on the short arm of rye chromosome 1 in bread wheat. Theor Appl Genet 119:783–793

Sharma S, Xu S, Ehdaie B, Hoops A, Close TJ, Lukaszewski AJ, Waines JG (2011) Dissection of QTL effects for root traits using a chromosome arm-specific mapping population in bread wheat. Theor Appl Genet 122:759–769

Villareal RL, Bañuelos O, Mujeeb-Kazi A, Rajaram S (1998) Agronomic performance of chromosomes 1B and T1BL. 1RS near-isolines in the spring bread wheat Seri M82. Euphytica 103:195–202

Waines JG, Ehdaie B (2007) Domestication and crop physiology: roots of green-revolution wheat. Ann Bot 100:991–998

Zeller FJ (1973) 1B/1R wheat-rye chromosome substitutions and translocations. In: Sears ER, Sears LMS (eds) Proceedings of the 4th international wheat genetics symposium. Columbia, MO, Missouri Agricultural Experiment Station, Columbia, 6–11 Aug 1973, pp 209–221

Effectiveness of Combination of Antibiotics on Different Isolates of '*Ralstonia solanacearum*'—A Dreaded Soil Born Phytopathogen and A Causative Agent of Bacterial Wilt

Rupa Verma, Abhijit Dutta, Ashok Kumar Choudhary and Sudarshan Maurya

Abstract

Few antibiotics have been found to be exhibiting antibacterial efficacy on different isolates of *Ralstonia solanacearum*, e.g., brinjal, capsicum, and tomato, causes bacterial wilt in enormous crops in tropical, subtropical and temperate regions across the world. In vitro, various combinations of antibiotics, i.e., ceftriaxone, gentamycin, and ambistryn in 1:1, 3:1, 1:3 ratios were screened and compared with application of antibiotics in their single form by agar well diffusion method. The zone of inhibition (ZOI) results reflected synergistic effect of combination of antibiotics on all three strains. The best result was observed with the combination of ceftriaxone–ambistryn (streptomycin) in 1:3 ratios, in Brinjal isolates. In capsicum, 1:3 ratio of the above antibiotics combination were either indifferent or antagonistic. In tomato strain, the combination of gentamycin and ambistryn in 1:3 ratios showed maximum efficacy. These results demonstrated indifferent effects for all combinations. The mode of action of combination therapy significantly differed from that of the same antibiotic acting individually; therefore, the selection of an appropriate combination of antibiotics are of paramount importance. It is expected that the combination therapy or synergistic therapy on this bacterial disease may be a step in the right direction for treating the wilt diseases in solanaceous plants to avoid antibiotic resistance.

R. Verma (✉) · A.K. Choudhary
Department of Botany, Ranchi University, Ranchi 834001, Jharkhand, India
e-mail: rupavermabiotech@gmail.com

A. Dutta
Department of Zoology, Ranchi University, Ranchi 834001, Jharkhand, India

S. Maurya
ICAR–Research Complex for Eastern Region, Research Centre,
Plandu, Ranchi 834010, Jharkhand, India

K. Mukhopadhyay et al. (eds.), *Applications of Biotechnology for Sustainable Development*, DOI 10.1007/978-981-10-5538-6_22

Keywords
Bacterial wilt · Combination of antibiotics · *Ralstonia solanacearum* · Synergistic effect of antibiotics

Introduction

Bacterial wilt disease, which has been impacting important solanaceous crops, e.g., tomato, potato, egg-plant, chilli, etc., solanaceous weeds like *Solanum nigram* and *Solanum dulcamara* (Martin and French 1985; French 1994), and some non-solanaceous crop, e.g., banana and groundnut both in tropical, subtropical, and temperate regions across the world including the eastern plateau and hill regions of India, where soil is mostly acidic (Anuratha et al. 1990) and causing enormous crop loss in these economically important crops, is the result of a bacterial pathogen *Ralstonia solnacearum* as diagnosed and proved by Smith first. Formerly known as *Psedomonas solanacearum*, bacterial pathogen *R. solnacearum*, which is an important soil-borne and devastating bacterial phytopathogens, aerobic in nature, non-spore former, Gram-negative, motile by a polar flagellar tuft, and the causative agent of bacterial wilt of solanaceous crops. The disease is also known as southern blight and *R. solanaceous* wilt and by other names at the various places of its occurrence (Kelman et al. 1954). Losses caused by the disease are known to be enormous but cannot be accurately estimated because abandonment of wilt susceptible crops in many parts of the world. Infected soil and surface water, including irrigation water are the primary sources of inoculum. The pathogen infects roots of susceptible plant, usually through wounds (Pradhanang et al. 2005), colonizes the xylem, prevents water movement into upper portion of plant tissues (Kelman 1954), and causes bacterial wilt in a very wide range of host plants (Agrios 2008). As the disease develops, all leaves may wilt quickly and desiccate although they remain green.

Biovars of *R. solanacearum* were differentiated according to their ability to oxidize disaccharides and hexose alcohol. (Hayward 1964). The races of *R. solanacearum* were identify by pathogenicity test on wide host range (Denny and Hayward 2001). The species of this bacterium was classified into five races on the basis of different host range (Buddenhagen et al. 1962) and six biovars according to the ability to oxidize disaccharides and three sugar alcohols (Hayward 1964; French 1994). In the following year, five biovars (based on carbon utilization pattern) and five pathogenic races were identified. Race1 occurs in the lowland tropics and warm temperate lands (French 1994).

Control of infection has been a major challenge due limited possibility for their chemical control, higher capacity of their survival in diverse environments, their high variability and their existence with an extremely wide host range (Nguyen and Ranamukhaarachchi 2010). Aqueous extracts of some plants having medicinal properties have been utilized in the past to control the disease caused by *R. solanacearum* but they have not been found to have significant inhibitory effects on the growth of *R. solanacearum* (Sangoyomi et al. 2011). Few antibiotics and peptides have also been used in the past to get rid of bacterial wilts. The most commonly used chemical treatment has been fumigation of contaminated soil/portion of farm with methyl bromide (Champoiseau et al. 2010) which is not only very expensive and tedious but applying these on large areas is difficult. Sodium hypochlorite is the another appropriate product which is used at field level for spot treatment of the holes left behind after rogueing of the wilting plants and for general field sanitation but its use is also expensive and tedious.

Despite a number of studies on control strategies of bacterial wilt by means of plant resistance and cropping system, a rigorous and complete protocol to control of the disease caused by *R. solanacearum* in various geographical regions is still lacking (Dalal et al. 1999). Hence, it is imperative to develop various antimicrobial agents for managing and containing *R. solanacearum* other than solvent extract. Expression of virulence factors in *R. solanacearum* is generally controlled by a complex regulatory network that responds to a variety of environmental conditions, the presence of host cells, and bacterial density (Schell 2000).

Hence with an aim to develop effective antibacterial agent that leaves no residual effect, this study investigates possible synergistic interactions among few antibiotics (Ambistryn, Ceftriaxone, Gentamycin) and their combined effectiveness vis a vis their effectiveness while using them singly for a potent effective means against three isolates of *R. solanacearum* and to analyze their in vitro antibacterial potential, antibacterial activities of antibiotics were assayed using the agar well diffusion technique. As aminoglycosides and tetracycline interfere with essential steps of protein synthesis and as most of the antibiotics interact with ribosomal RNA (Yoshizawa et al. 2008), the ribosome happens to be the central target of many important antibiotics for inhibiting the protein synthesis of the bacterium and for consequently ceasing its growth and propagation.

Methods and Materials

Surveying, Identification and Collection of Infected Plants

A survey was carried out to understand the impact and severity of bacterial wilt of brinjal, tomato, and capsicum at some of the selected districts viz. Ranchi, Simdega, Khunti, Palamu, and Hazaribagh of Jhrakhand. Because of the incidence of bacterial infection, leaves of the infected plants were observed having turned to yellow and the area between leaf veins were dead and became brown although the plants remained upright. Collected stem segments, from collar region of wilted plants, were rinsed with sterilized distilled water containing 1% Clorox. Stem streaming methods, which is a valuable diagnostic tool for quick detection of bacterial wilt in the field (Allen et al. 2001) was used for collecting samples.

For a quick field diagnosis, the streaming of milky white masses of bacterial cells (ooze) confirmed the disease is bacterial wilt caused by *R. solanacearum* and to distinguish bacterial wilt from vascular wilts caused by fungal pathogen and nematode. Samples of the diseased plants were collected from each of the surveyed district and were brought to the laboratory for the isolation of different group of isolates of *R. solanacearum*. The symptoms of bacterial wilt were observed and labeled properly.

Isolation and Maintenance of *R. solanacearum* Isolates

Turbid bacterial suspensions of all three infected plants were streaked on the nutrient agar (NA) plates separately and incubated at 28 °C for not less than 24 h for growing the bacteria in the medium. After isolation, *R. solanacearum* isolates were purified by streaking a single colony of each isolate on triphenyl tetrazolium chloride (TTC) plate (Kelman 1954). These isolates of *R. solanacearum* were then preserved in King's B Media for its subsequent studies. The pathogenicity test was performed to confirm about the isolates of *R. solanacearum*, on brinjal, capsicum, and Tomato. A single colony of *R. solanacearum* showing virulent, fluidal, irregular, and creamy white with pink at the center was selected for each group of isolates. On the plates of brinjal and capsicum, this bacterial isolates were whitish in color. Isolations of tomato were golden yellow in color.

Preparation of Single and Combination of Antibiotic Stock Solutions

Three antibiotics viz ceftriaxone, ambistryn, and gentamycin were selected for using them singly and in combinations for screening against the isolations of the bacterium *R. solanacearum* from capsicum, brinjal, and tomato. Ceftriaxone is a known sterile, semisynthetic, broad spectrum, cephalosporin antibiotic that inhibits cell wall synthesis. Ambistryn contains streptomycin and is an aminoglycoside antibiotic (procured from Himedia Pharmaceuticals, India for use in experiments). The primary mechanism of action of streptomycin is to bind the bacterial ribosomes irreversibly and thereby inhibiting protein synthesis. Gentamicin is a broad spectrum antibiotic and irreversibly binds 30S subunit and inhibits protein synthesis. 500 ppm stock solution was prepared for each of the three antibiotics (ceftriaxone, ambistryn, and gentamycin). Different combinations 1:1, 3:1, and 1:3 ratios were prepared from stock solution of single antibiotics (Ceftriaxone, Gentamycin, Ambistryn). Three combinations that were used are ceftriaxon and ambistryn (C+A), ceftriaxone with gentamycin (C+G) and ambistryn with gentamycin (A+G).

Microbiological Screening by Agar Well Diffusion Method

Antimicrobial activity of different antibiotics were analyzed by the Agar well Diffusion method (Murray et al 1995, modified by Olurinola et al. 1996). This method was used for the comparison of the single dose of antibiotics with its various combinations, i.e., in 1:1, 1:3, and 3:1 ratios. Petri plates, test tubes, and Nutrient agar (Agar-16 g, nutrient broth-13 g, Distilled water-1000 ml) were sterilized at 121^{c} for 15 min and media was poured on all plates of each isolates and allowed to solidify. These plates were swabbed (sterile cotton swabs) with fresh grown bacterial culture of dilution (10^{-3}cfu/ml). Inoculums were allowed to dry for 5 min. By using 5 mm cork-borer, five wells were made in each of the 60 plates, one for control at the center and others for 10, 20, 30 and 40 µg mL^{-1} of antibiotics solution, for each selected three isolates of tomato, capsicum and brinjal. Sterile distilled water was used as control. With the help of micro pipette 10, 20, 30, and 40 µg $mL^{-1}$of antibiotic solutions were added in each of the 60 plates (having three isolations) and allowed to incubate for 24 h at 29–30 °C. Four sets of experiments were performed for each of the brinjal, capsicum, and tomato isolates of *R. solanacearum*. One set for single application of antibiotics and other three for three different combinations (1:1, 1:3, and 3:1) five replications were maintained for each set of experiments. The diameter of the zone of inhibition (in mm) of all twelve sets of experiment was measured. Data, thus obtained, was analyzed using the Statgraphics software.

Results and Discussion

Combinations of ceftriaxone and ambistryn, ceftriaxone and gentamycin, and ambistryn and gentamycin produced desired improved result in all of the four concentrations in the case of **brinjal** at 1:1 ratio. Hence rather than using these drugs singly, combinations was found to be more effective in producing better result as combinations of antibiotics, i.e., C+A, C+G, and A+G produced best zone of inhibition result, i.e., from 13.2 to 19.5 mm compared to 11.17 to 16.7 mm of ZOI at various concentrations when they were used singly. Combination of ceftriaxone and ambistryn, when mixed in the ratio of 3:1 and used on brinjal isolates of *R. solanacearum* had better impact than the individual performance of ceftriaxone and ambistryn at all the other concentrations. Combination of G+A and G+C produced almost similar impact that of individual effectiveness at all concentrations at the same ratio. When the chosen antibiotics viz ceftriaxone, gentamycin, and ambistryn were mixed in 1:3 ratio, all of the two possible combinations reflected strong efficacy at all concentrations, succeeded exhibiting ZOI ranging from 13.8 to 20.5 mm compared to the individual performance of chosen drugs, i.e., from 5.6 mm at lower concentration to 17.9 mm at higher concentrations. Both, gentamycin and ceftriaxone singly were producing individual ZOI of 5.6 and 9.45 mm at 10 and 20 ppm singly. But both of them, coupling together and used in

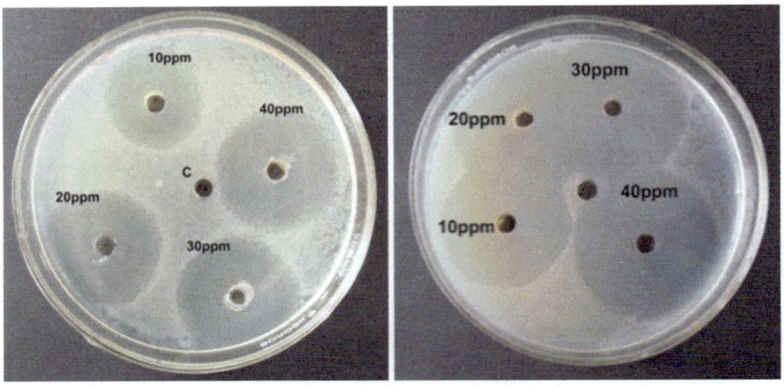

Fig. 1 Antimicrobial efficacy (showing ZOI) of antibiotics (in single and combinations 1:1 ratio), in brinjal isolates of *R. solanacearum* at 10, 20, 30 and 40 $\mu g\ mL^{-1}$ or ppm

Graph 1 Average ZOI (mm), as reflected by the single and combinations of antibiotics on brinjal isolates of *R. solanacearum* at different concentrations

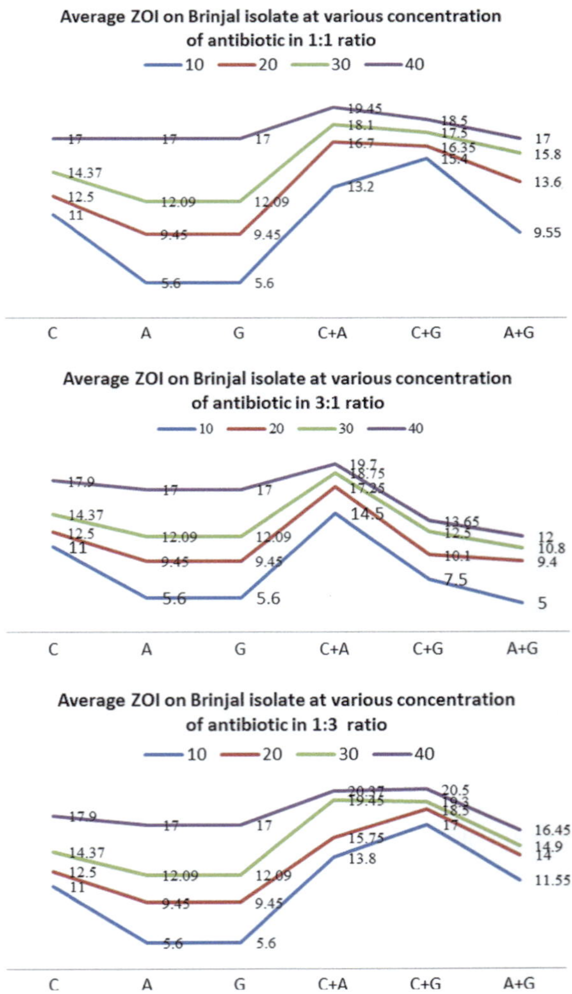

combination at 1:1 ratio, resulted in a ZOI of 13.8 and 15.75 mm. ZOI even increased to 19.45 mm in the same combination from 12.09 mm compared to single use of drugs (Fig. 1; Graph 1).

In the case of **capsicum**, ceftriaxone alone was effective in containing the bacteria at all concentrations and showed best result when used singly rather than in combinations. Zone of Inhibition showed by ceftriaxone in combination of ambistryn or gentamycin was observed to be less effective than C as alone. Combination of ambistryn and gentamycin had insignificant result, when used in the ratio of 1:1. Gentamycin in combination with ceftriaxone had better result than that of gentamycin alone at 1: 1 ratio. Using Ambistryn alone was observed to be better than using the combination of C+A at the similar concentration at 1:3 ratios. Zone of inhibition of ceftriaxone, when used singly, was observed as much higher (5.6–17 mm) than the ceftriaxone coupled either with ambistryn or gentamycin in the ratios of 1:3 (ZOI from 1.6 to 4.2 mm). But A+G produced significant and marked improved result with ZOI of 11.9–6.8 mm of ZOI at various increasing concentrations. At 3:1 ratio, none of the combination of antibiotics produced desired result and rather lost even their individual potency and hence was observed as unworkable proposition (Fig. 2; Graph 2).

In the case of **tomato**, combination of Gentamycin with ceftriaxone produced better result of ZOI than the single use of gentamycin at all concentrations at 1:1 ratio. Rest of the other combinations, i.e., but C+A was observed as yielding same result that of ceftriaxone and ambistryn, when used on a standalone basis. Gentamycin yields best result when used in combination with ceftriaxone at all concentrations at 3:1 ratio and it was observed to be as best alternative to single use of gentamycin. Mixing of ambistryn with gentamycin at all concentrations did not yield positive result and antibiotics, rather, lost their individual potency. Combination with ambistryn with ceftriaxone produces best result in containing the bacterial infection, when used in 1:3 ratios and at all the used concentrations.

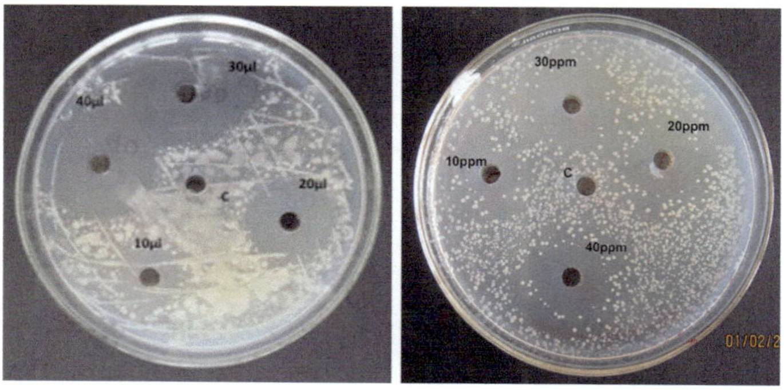

Ceftriaxon(Single) C+G1:1(combination)

Fig. 2 Antimicrobial efficacy (showing ZOI) of antibiotics (in single and combinations in 1:1 ratio), in capsicum isolates at 10, 20, 30 and 40 µg mL^{-1} or ppm

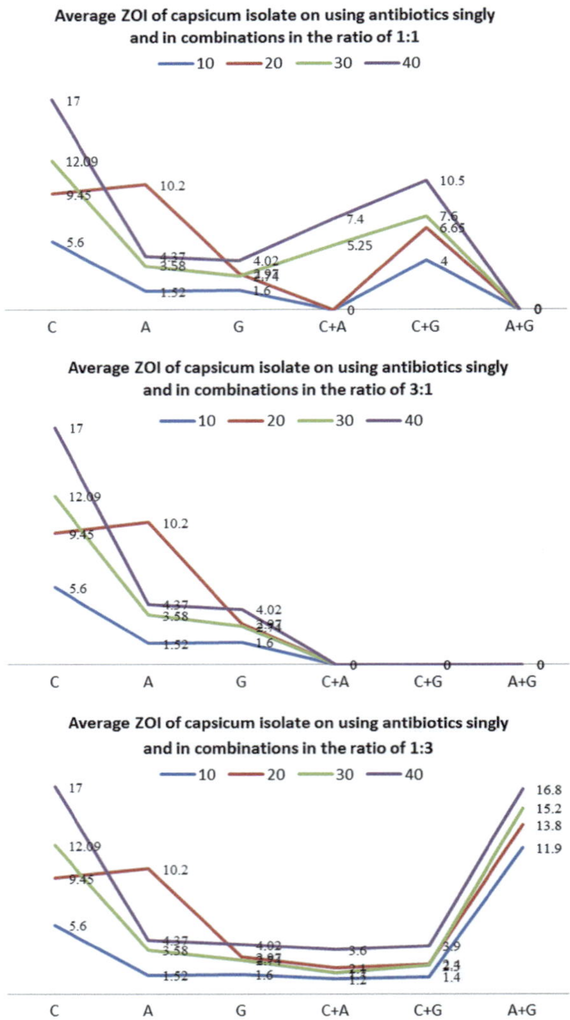

Graph 2 Average ZOI (mm), as reflected by the single and combinations of antibiotics on capsicum isolates of *R. solanacearum* at different concentrations

Hence, rather than using gentamycin singly, the combination of ambistryn and ceftriaxone may be considered as best option (Fig. 3; Graph 3).

Statistical Analysis

Captioned tables show various statistics for each of the three columns of data with different combinations of drug. To test for significant differences amongst the column means, Analysis of Variance (ANOVA) was performed. The ANOVA table decomposes the variance of the data into two components: a between-group component and a within-group component. The F-ratio is a ratio of the

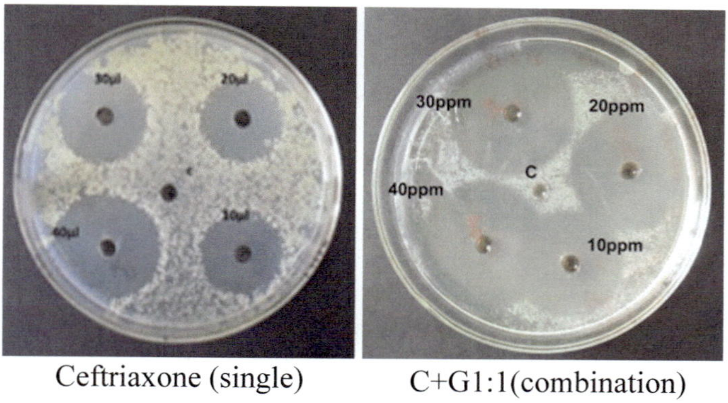

Fig. 3 Antimicrobial efficacy (showing ZOI) of antibiotics (in single and combinations in 1:1 ratio), in tomato isolates at 10, 20, 30 and 40 μg mL^{-1} or ppm

Graph 3 Average ZOI (mm), as reflected by the single and combinations of antibiotics on tomato isolates of *R. solanacearum* at different concentrations

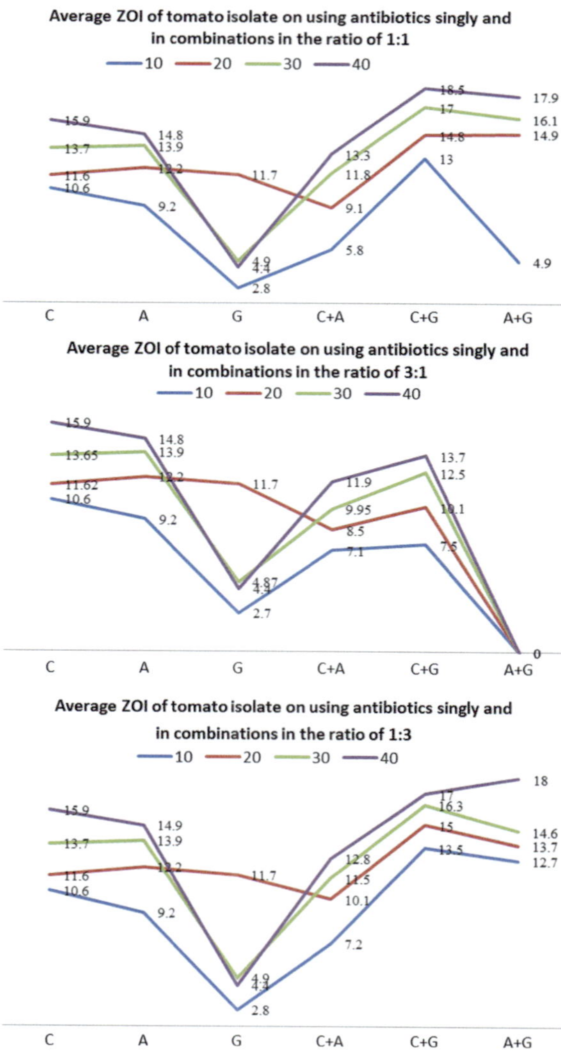

between-group estimate to the within-group estimate. If the P-value of the F-test is greater than or equal to 0.05, there is not a statistically significant difference between the means of the 3 variables at the 95.0% confidence level.

Conclusion

Antibiotics and its various combinations at different ratios (1:1, 1:3, 3:1) showed antibacterial property against three isolates of *R. solanacearum* in brinjal, capsicum, and tomato isolates. Antibiotics used in combination were observed to be highly to moderately effective, when used in selected ratio and appropriate concentrations and inhibits the growth of bacterial infection on the chosen isolates of *R. solanacearum*. But, in few cases, use of antibiotics alone was observed to be potent enough compared to synergistic impact produced by their combination which was almost observed as negative or zero. So the combinations of antibiotics can be the main basis of wilt control in order to avoid antibiotic resistance.

Acknowledgements We extend our sincere thanks to the director, ICAR-Research Complex for Eastern Region, Research Center, Ranchi, India, for being kind in providing the necessary laboratory facilities to carry out this research.

References

Agrios GN (2008) Plant pathology. Elsevier Academic Press, Amsterdam, pp 647–649

Allen C, Kelman A, French ER (2001) Brown rot of potatoes. In: Stevenson WR, Loria R, France GD, Weingarten DP (eds) Compenclium of potato disease, 2nd edn. APS Press, St. Paul, MN, pp 11–13

Anuratha CS, Gnanamanickam SS (1990) Biological control of bacterial wilt caused by *Pseudomonas solanacearum* in India with antagonistic bacteria. Plant Soil 124:109–116

Buddenhagen I, Sequeira L, Kelman A (1962) Designation of races in *Pseudomonas solanacearum*. Phytopathology 52:726

Champoiseau PG, Jones JB, Momol TM, Pingsheng J, Allen C, Norman DJ, Caldwell K, Chu W (2010) Laboratory guide for identification of plant pathogenic bacteria, 3rd edn. APS Press, st. Paul, MN, pp 151–174

Dalal NR et al (1999) Studies on grading and prepackaging of some bacterial wilt resistant brinjal (*Solanummelogena* L.) varieties. J Soils Crops 9(2):223–226

Denny TP, Hayward AC (2001) Gram negative bacteria; Ralstonia. In Schaad NW, Jones JB (eds)

French ER (1994) Strategies for integrated control of bacterial wilt of potatoes. In: Hayward AC, Hartman GL (eds) Bacterial wilt: the disease and its causative agents, *Psedomonas solanacearum*. CAB International, Wallingford, pp 98–113

Hayward AC (1964) Characteristic of *Pseudomonas solnacearum*. J Appl Bacteriole 27:265–277

Kelman A (1954) The relationship of pathogenicity in *Pseudomonas solanacearum* to colony appearance on atetrazolium medium. Phytopathology 44:693–695

Martin C, French ER (1985) Bacterial wilt of potatoes caused by *Psedomonas solanacearum*. CIP Tech Inf Bull 13:1–6 (CIP, Lima Peru)

Murray PR, Baron EJ, Pfaller MA, Tenover FC, Yolken HR (1995) Manual of clinical microbiology, 6th edn. ASM Press, Washington, DC, pp 15–18

Nguyen MT, Ranamukhaarachchi SL (2010) Soil-borne antagonists for biological control of bacterial wilt disease caused by *Ralstonia solanacearum* in tomato and pepper. J Plant Pathol 92(2):395–406

Pradhanang PM et al (2005) Application of acibenzolar-S-methyl enhances host resistance in tomato against *Ralstonia solanacearum*. Plantdisease 74:13–17

Ralstonia solanacearum Race 3 biovar2 causing brown rot of potato, Bacterial wilt of tomato and southernwilt of geranium (online). Available at http://plantpath.ifas.ufl.edu/rsol/NRI-project/projectsummary.html. Accessed 25 June 2010. American Phytopathological Society, Madison, WI

Sangoyomi TE, Oweseni AA, Adebayo OS, Omilani OA (2011) Evaluation of some botanicals against bacterial wilt of tomatoes. Int Res J Microbiol 2(9):365–369

Schell MA (2000) Control of virulence and pathogenicity genes of *Ralstonia solanacearum* by an elaborate sensory network. Annu Rev Phytopathol 38:263–292

Yoshizawa1 S, Fourmy D, Puglisi JD (2008) Department of Structural Biology, Stanford University School of Medicine, Stanford, CA

Modelling of L-protein from Ebola Virus and Development of Its New Inhibitor Molecules: An In Silico Approach

Ekta, Shubham Choudhury, Priyam Rout, Santosh Kumar, Pravin Kumar and Raju Poddar

Abstract

L-protein is an RNA-dependent RNA polymerase. It creates many new copies of RNA genome. It is one of the potential and promising anti-Ebola drug targets. It is essentially required for the growth, survival and pathogenesis of the Ebola virus. Thus, to find potent anti-Ebola agents, we have modeled three-dimensional structure of L-protein. The modeled was validated through different tools and subsequently energy minimized. Then structure of potential inhibitor molecules was predicted through pharmacophore models which are based on the existing potent inhibitors' binding modes. Molecular dynamics simulation and docking studies were performed. New potent molecule was reported. Further, the stability of protein–drug complex was also executed. This proposed model and docking study would play a vital role to identify potential lead compounds and treatment against L-protein of Ebola virus.

Keywords

Ebola virus · L-protein · Molecular dynamics simulation · Docking · RNA polymerase

Ekta · S. Choudhury · P. Rout · S. Kumar · P. Kumar · R. Poddar (✉)
Department of Bio-Engineering, Birla Institute of Technology,
Mesra, Ranchi 835215, Jharkhand, India
e-mail: rpoddar@bitmesra.ac.in

K. Mukhopadhyay et al. (eds.), *Applications of Biotechnology for Sustainable Development*, DOI 10.1007/978-981-10-5538-6_23

Introduction

Ebola virus is a single-stranded RNA virus consisting of threadlike filaments. The Ebola virus belongs to a family of viruses called *Filoviridae* (Kuhn et al. 2010). The viral replication is performed by viral RNA-dependent RNA polymerase only. Infection with a specific species causes the Ebola virus disease (EVD), which was earlier known as Ebola hemorrhagic fever. The symptoms of the disease usually start with fever along with fatigue, joint and muscular ache with a mortality rate of about 59% (Dixon and Schafer 2014). Currently, there are very limited treatment options for Ebola virus disease (EVD). A lot of research and development is taking place to develop treatments as well as to create a vaccine to prevent EVD (Richardson et al. 2010). Current treatment for Ebola virus involves experimental drugs like Zmapp, BCX4430, Brincidofovir, DZNep, Favipiravir and TKM-Ebola which are under study (Huggins et al. 1999; Kanapathipillai et al. 2014). There are two types of vaccines for Ebola virus namely cAd3-EBOV and VSV-EBOV are in advanced stages of approval (Pavot et al. 2016; Dufe et al. 2007).

In this paper, we attempt to predict and validate the tertiary structure of RNA Polymerase of Ebola and Design new drugs/inhibitor molecules for RNA Polymerase of Ebola Virus using bioinformatics tools, servers and software.

Methodology

The amino acid sequence of L-protein, partial [Zaire Ebola virus] having accession Number: AIN75250.1, was downloaded from NCBI protein database (available at: https://www.ncbi.nlm.nih.gov/protein) for homology modelling. To generate a model, the coordinates of template and sequence alignment file was provided to MODELLER 9.12 software (Laskowski et al. 1993; Sali et al. 1995) as an input. Five such models were generated and ranked based on their Discrete Optimized Protein Energy (DOPE) score (van Gunsteren et al. 1996).

Molecular dynamic simulation study of all systems (of the predicted structure) was carried out with the help of GROMACS 4.5.6 software package (William L. Jorgensen) by treating protein with OPLS-AA/L aal atom force field (Berendsen et al. 1981) for preparing the topology file. The system was solvated with the charged protein. Genion tool helps to replace water molecules with ions. The entire system was subjected to energy minimization by steepest descent algorithm for 50,000 steps and acceptance limit of 1000 kJ mol^{-1} nm^{1}. Further, two steps equilibrium at 300 K temperature was adapted for the system, first, under constant number of particles, Volume and Temperature (NVT) ensemble was implemented for 100 ps using v-rescale method for temperature coupling. Then, for pressure coupling constant number of particles, Pressure and Temperature, i.e. isothermal-isobaric ensemble was applied for 500 ps with Parrinello–Rahman method (Bussi et al. 2007). Upon completion of these two equilibration phases, the system was well-equilibrated at the desired temperature and pressure. Several inhibitors have been investigated from various literatures. The chemical structure of these inhibitors was drawn using chemsketch (https://www.acdlabs.com/resources/freeware/chemsketch/) and converted into .pdf and .sdf format for further use. Twenty different conformations of inhibitors were generated and subjected for docking against simulated structure by using MAESTRO software.

Results

Using NCBI protein database, target amino acid sequence of L-protein, partial [Zaire Ebolavirus] Accession No-AIN75250.1 was retrieved for protein modelling. BLASTP search of L-protein amino acid sequence against PDB database identified the domain of fyn protein from Mus Musculus (as the most appropriate template having query coverage of 35%. The monomer tertiary structure of L-protein was generated by MODELLER 9.12 is shown in Fig. 1a. The crystal structure of RNA Polymerase of Ebolavirus was predicted and was solved at a resolution of 1.9 Å. The modelled tertiary structure (Fig. 1a) showed favourable Ramachandran value of 83.50% (red region), additional allowed region 10.40% (brown region), generously allowed region 2.5% (yellow region) and disallowed region 1.8% (pale yellow region).

The validation of modelled structure through MD Simulation inferred that there were fewer fluctuations in RMSD when the protein was simulated for 10 ns (Fig. 1b, c). Figure 1b gives the energy evolution of the protein. In Fig. 1c graph shows that at the beginning of simulation, RMSD value of protein structure had fluctuated up to 6 ns but after that it became constant. From the figure, it can be inferred that overall potential energy of RNA Polymerase is in a range of −2,380,000 to −2,370,000. It reveals that the modelled structure is energy minimized and this can be used for further research. A number of inhibitors, such as 3-(2-carboxy-ethyl)-4, 6-dichloro-1 h-indole-2-carboxylic acid have been searched by literature survey and Drug Bank. Its chemical structure in pdf format and sdf format has been retrieved from chemsketch (https://www.acdlabs.com/resources/freeware/

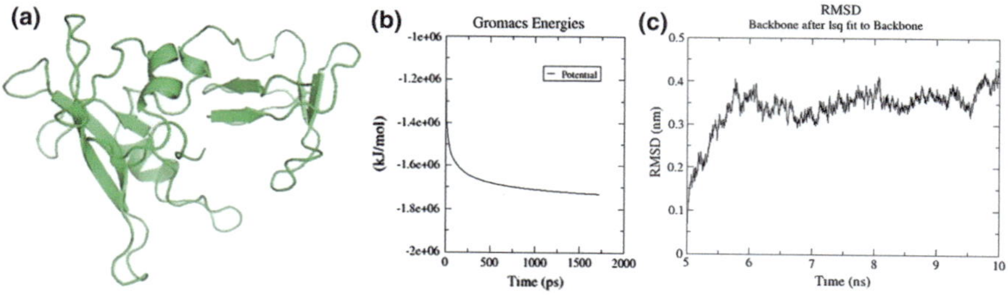

Fig. 1 **a** Modelled tertiary structure of RNA polymerase of ebola virus (using MODELLER 9.12), **b** Time evolution graph representing potential energy, and **c** Time evolution for RMSD of the predicted structure

chemsketch/). Finally, following molecules were selected for docking. FT-site finder application revealed that there were 3 active sites. Site 1 consists of residues TYR71, GLU99, ARG100, THR101, CYS102, TYR114, ASN115, HIS118, LYS119, PHE159, SER160, LEU161 and LYS164. Twenty different conformations of each inhibitor have been constructed. Inhibitors are subjected for docking against simulated structure by using MAESTRO software. 3D structures of the target protein (L-protein) and the ligand were docked using MAESTRO, prepared by LIGPREP with default parameters.

All the designed analogs have been docked with simulated structure and it was found that drug1 has good binding affinity among the docked analogs (Fig. 2). Most of the ligand were bound to the conserved residues Lys161, Ser160, Tyr114, Lys119, Asn115, Thr101. Pockets of RNA polymerase contains, Phe159, Lys164,

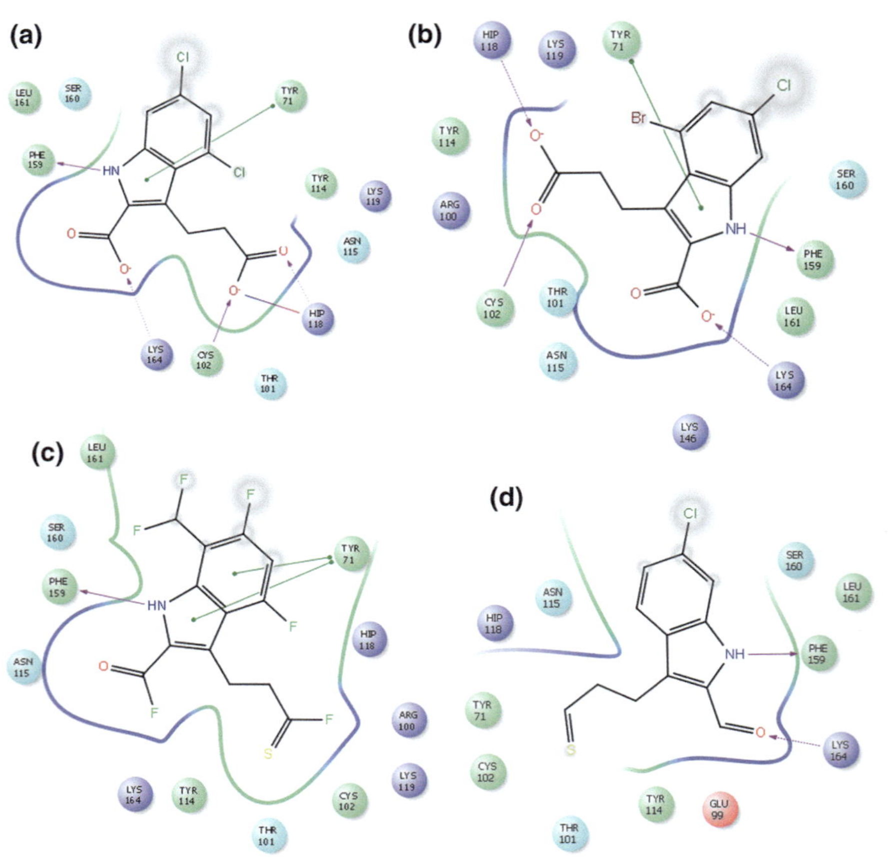

Fig. 2 Interactions between four Protein and drug 1 (**a**), inhibitor molecule 2 (**b**), inhibitor molecule 3 (**c**), and inhibitor molecule 4 (**d**)

Table 1 Docking score for different drug molecules with target molecule

Drug No.	Site 1	Site 2	Site 3
Drug 1	−6.72241	−5.17132	−5.17969
Drug 2	−6.12683	−4.96707	−4.35746
Drug 3	−5.19904	−4.84837	−4.255

Tyr71, Hip118, Cys102. The interaction of inhibitors with these residues ensures inhibition of enzyme activity. The potential inhibitors identified by current work may be used as a lead for drug designing against Ebola virus. Table 1 represents the docking score at differ sites of the target.

Conclusion

The current study was carried out to develop and validate novel drug molecules against L-protein of Ebola virus. It will lead to design and synthesis more specific novel compounds as antiviral agents. The docking, results of designed ligand analogs revealed that these analogs would be a novel scaffold in the design. Drug 1, namely [IUPAC: OC(=O)CCC1C(NC2=C1C(=CC(=C2)Cl)Cl)C(O)=O] shows better affinity towards L-protein and will be ideal for antiviral agent specially to target the inhibition of Ebola virus.

References

Berendsen HJC, Postma JPM, van Gunsteren WF, Hermans J (1981) Interaction models for water in relation to protein hydration, In: Pullman B (ed) Intermolecular forces, D. Reidel Publishing Company, Dordrecht, pp 331–342

Bussi G, Donadio D, Parrinello M (2007) Canonical sampling through velocity rescaling. J Chem Phys 126(1):014101

Dixon MG, Schafer IJ (2014) Ebola viral disease outbreak-west Africa. MMWR Morb Mortal Wkly Rep 63(25):548–551

Dufe VT, Ingner D, Heby O, Khomutov AR, Persson L, Al-Karadaghi S (2007) A structural insight into the inhibition of human Ebola virus disease by 1-amino-oxy-3-aminopropane. Biochem J 405:261–268

Huggins J, Zhang ZX, Bray M (1999) Antiviral drug therapy of filovirus infections: S-adenosylhomocysteine hydrolase inhibitors inhibit Ebola virus in vitro and in a lethal mouse model. J Infect Dis 179:240–247

Kanapathipillai R et al (2014) Ebola vaccine—an urgent international priority. N Engl J Med 371 (24):2249–2256

Kuhn JH et al (2010) Proposal for a revised taxonomy of the family Filoviridae: classification, names of taxa and viruses, and virus abbreviations. Arch Virol 155(12):2083–2103

Laskowski RA, MacArthur MW, Moss DS, Thornton JM (1993) PROCHECK: a program to check stereo chemical quality of protein structures. J Appl Crystallogr 26:283–291

Pavot V (2016) Leading Ebola vaccine candidates. Vaccin Res Open J 1(1):60–71

Richardson Jason S et al (2010) Recent advances in Ebolavirus vaccine development. Hum Vaccines 6(6):439–449

Sali A, Pottertone L, Yuan F, Van Vlijmen H, Karplus M (1995) Proteins evaluation of comparative protein modelling by MODELLER. Proteins 23(3):318–326

van Gunsteren WF, Billeter SR, Eising AA, Hünenberger PH, Krüger P, Mark AE, Scott WRP, Tironi IG (1996) Biomolecular simulation: The GROMOS96 manual and user guide. Verlag der Fachvereine Hochschulverlag, AG an der ETH Zurich, Groningen

MIX
Papier aus verantwortungsvollen Quellen
Paper from responsible sources
FSC® C105338

If you have any concerns about our products,
you can contact us on
ProductSafety@springernature.com

In case Publisher is established outside the EU,
the EU authorized representative is:
Springer Nature Customer Service Center GmbH
Europaplatz 3, 69115 Heidelberg, Germany

Printed by Libri Plureos GmbH
in Hamburg, Germany